Lecture Notes of the Institute for Computer Sciences, Social Informatics and Telecommunications Engineering 383

More information about this series at http://www.springer.com/series/8197

Navin Kumar · M. Vinodhini ·
Ranga Rao Venkatesha Prasad (Eds.)

Ubiquitous Communications and Network Computing

4th EAI International Conference, UBICNET 2021
Virtual Event, March 2021
Proceedings

 Springer

Editors
Navin Kumar (iD)
Amrita School of Engineering
Bangalore, India

M. Vinodhini (iD)
Amrita School of Engineering
Bangalore, India

Ranga Rao Venkatesha Prasad
Software Technology
Delft University of Technology
Delft, The Netherlands

ISSN 1867-8211   ISSN 1867-822X (electronic)
Lecture Notes of the Institute for Computer Sciences, Social Informatics
and Telecommunications Engineering
ISBN 978-3-030-79275-6   ISBN 978-3-030-79276-3 (eBook)
https://doi.org/10.1007/978-3-030-79276-3

This Springer imprint is published by the registered company Springer Nature Switzerland AG
The registered company address is: Gewerbestrasse 11, 6330 Cham, Switzerland

Preface

We are delighted to introduce the proceedings of the 4th edition of the European Alliance for Innovation (EAI) International Conference on UBIquitous Communications and NETwork Computing (UBICNET 2021). Even during the time of the COVID-19 pandemic, this conference brought together researchers, developers, and practitioners on one platform to discuss the advances in various fields of communication, like 5G and interconnected systems. The theme of the conference was "Intelligent and Cognitive Connectivity."

The technical program of UBICNET 2021 was highly selective, comprising 17 full papers given as oral presentation in the main conference track. The track was arranged in different sessions: 5G Networks, Millimeter Wave Communication Systems, and Emerging Applications; Quantum Communication, IoT, and Emerging Applications; Data Analytics and Cloud Computing; and Artificial Neural Networks, Machine Learning, and Emerging Applications. Besides the high-quality technical paper presentations, the technical program also featured three keynote speeches. Excellent keynotes speeches by experts from the industry focused on highly complex technology innovation. Many challenges and associated opportunities in the next generation networks were discussed.

We believe that, despite the challenges of the pandemic, the conference was highly successful. The success was due to the well-structured coordination with the steering chair, Imrich Chlamtac Dr. T K Ramesh helped us to compile the high-quality technical program. The Technical Program Committee co-chair, Dr. Abhilash Ravikumar, played an important role in getting papers reviewed on time. Special mention should be made of our local chairs, Prof. Sagar B. and Vignesh V., who ensured every requirement of the conference was properly arranged. The conference management and EAI staff were quick in responding to the queries, which is another reason for the success of the conference. We sincerely appreciate their constant support and guidance. It was also a great pleasure to work with such an excellent Organizing Committee who worked very hard in organizing and supporting the conference, and we are particularly thankful for the efforts of the Technical Program Committee. We are grateful to the conference manager, Ms. Natasha Onofrei, and Radka for their continuous support. In addition, we are very grateful to all the authors who submitted their papers to UBICNET 2021.

We strongly believe that the UBICNET conference provided a good forum for all researchers, developers, and practitioners to discuss the relevant issues related to technology, research, and development. We are sure that future editions of the UBICNET conference will be as successful and stimulating as indicated by the contributions presented in this volume.

April 2021

Navin Kumar
M. Vinodini
R. Venkatesha Prasad

Organization

Steering Committee

Imrich Chlamtac — University of Trento, Italy
Navin Kumar — Amrita School of Engineering Bangalore, Amrita Vishwa Vidyapeetham, India

Organizing Committee

General Chair

T. K. Ramesh — Amrita School of Engineering Bangalore, Amrita Vishwa Vidyapeetham, India

R. Venkatesha Prasad — Delft University of Technology, the Netherlands

Technical Program Committee Chair and Co-chairs

Abhilash Ravikumar — Amrita School of Engineering Bangalore, Amrita Vishwa Vidyapeetham, India

Qui Li — Kent University, UK
Sheeba Kumari M. — New Horizon College of Engineering, India

Sponsorship and Exhibit Chair

Kirti S. Pande — Amrita School of Engineering Bangalore, Amrita Vishwa Vidyapeetham, India

Local Chair

Sagar B. — Amrita School of Engineering Bangalore, Amrita Vishwa Vidyapeetham, India

Publicity and Social Media Chair

Vignesh V. — Amrita School of Engineering Bangalore, Amrita Vishwa Vidyapeetham, India

Aaryan Oberoi — Amrita School of Engineering Bangalore, Amrita Vishwa Vidyapeetham, India

Publications Chairs

M. Vinodhini — Amrita School of Engineering Bangalore, Amrita Vishwa Vidyapeetham, India

Reema Sharma — Visvesvaraya Technological University, India

Web Chair

P. Sathish Kumar   Amrita School of Engineering Bangalore, Amrita
                                Vishwa Vidyapeetham, India

Posters and PhD Track Chair

Jalpa Shah   Amrita School of Engineering Bangalore, Amrita
                                Vishwa Vidyapeetham, India

Panels Chair

Sunitha R.   Amrita School of Engineering Bangalore, Amrita
                                Vishwa Vidyapeetham, India

Demos Chair

Lalitha S.   Amrita School of Engineering Bangalore, Amrita
                                Vishwa Vidyapeetham, India

Tutorials Chair

Priya B. K.   Amrita School of Engineering Bangalore, Amrita
                                Vishwa Vidyapeetham, India

Technical Program Committee

Abhilash Ravikumar   Amrita Vishwa Vidyapeetham, India
K. Navin Kumar   Amrita Vishwa Vidyapeetham, India
T. K. Ramesh   Amrita Vishwa Vidyapeetham, India
Anand M.   Centre for Development of Telematics, India
Priya B. K.   Amrita Vishwa Vidyapeetham, India
Chinthala Ramesh   Amrita Vishwa Vidyapeetham, India
Kamatchi S.   Amrita Vishwa Vidyapeetham, India
Kaustav Bhowmick   Amrita Vishwa Vidyapeetham, India
Lalitha Mohan   BIT, India
Mahesh Kumar Jha   CMR Institute of Technology, Bengaluru, India
M. Vinodhini   Amrita Vishwa Vidyapeetham, India
Monika Soni   All India Council for Technical Education, India
Manik Sharma   DAV University, India
Nandi Vardhan H. R.   Amrita Vishwa Vidyapeetham, India
Satish Kumar P.   Amrita Vishwa Vidyapeetham, India
Prakash G.   Amrita Vishwa Vidyapeetham, India
Sagar B.   Amrita Vishwa Vidyapeetham, India
Soma Pandey   Ramaiah University of Applied Sciences, India
Reema Sharma   Entrepreneur, India
Sheeba Kumari M.   New Horizon College of Engineering, India
Singh Monika   CMR Institute of Technology, Bengaluru, India
Shachi P.   BMS College of Engineering, India

Sreeja Kochuvila	Amrita Vishwa Vidyapeetham, India
Sreeja Sukumaran	Christ University, India
Subramanian S.	Intel, Bangalore, India
Udaya Sankar V.	SRM University, India
Vignesh V.	Amrita Vishwa Vidyapeetham, India
Sanjeev G.	PES University, India
Srikanth K. S.	Intel, Bangalore
Ganapathy Hegde	Amrita Vishwa Vidyapeetham, India
Aravinda K.	New Horizon College of Engineering, India
Jaya R.	New Horizon College of Engineering, India
Giriraja C. V.	Amrita Vishwa Vidyapeetham, India
Bhavana V.	Amrita Vishwa Vidyapeetham, India
Kannadhasan S.	Cheran College of Engineering, India
Jayashree Oli	Amrita Vishwa Vidyapeetham, India
Jalpa Shah	Amrita Vishwa Vidyapeetham, India
Suma M. N.	BMS College of Engineering, India
Neelima N.	Amrita Vishwa Vidyapeetham, India
Rakesh N.	Amrita Vishwa Vidyapeetham, India
Sanjika Devi	Amrita Vishwa Vidyapeetham, India
Sonali Agarwal	Amrita Vishwa Vidyapeetham, India
Lalitha S.	Amrita Vishwa Vidyapeetham, India
Uma Maheshwari	Amrita Vishwa Vidyapeetham, India
P. Maran	Amrita Vishwa Vidyapeetham, India
Paramashivam C.	Amrita Vishwa Vidyapeetham, India

Contents

5G Networks, Millimeter Wave Communication Systems and Emerging Applications

Analysis of Generic Routing Encapsulation (GRE) over IP Security (IPSec) VPN Tunneling in IPv6 Network 3
  Md. Raihan Uddin, Nawshad Ahmad Evan, Md Raiyan Alam, and Md. Taslim Arefin

Mobility Management Based Mode Selection for the Next Generation Network 16
  Pallavi Sapkale, Uttam D. Kolekar, and Navin Kumar

Specification of a Framework, Fully Distributed, for the Management of All Types of Data and the Services Close to Users 26
  Thierno Ahmadou Diallo

A Leading Edge Detection Algorithm for UWB Based Indoor Positioning Systems. 45
  Sreenivasulu Pala and Dhanesh G. Kurup

Designing Multiband Millimeter Wave Antenna for 5G and Beyond 56
  Tirumalasetty Sri Sai Apoorva and Navin Kumar

Analog Beamforming mm-Wave Two User Non-Orthogonal Multiple Access 66
  S. Sumathi, T. K. Ramesh, and Zhiguo Ding

Quantum Communication, IoT and Emerging Applications

A Novel Multi-User Quantum Communication System Using CDMA and Quantum Fourier Transform 79
  M. Anand and Pawan Tej Kolusu

Qubit Share Multiple Access Scheme (QSMA). 91
  Pawan Tej Kolusu and M. Anand

Design of CoAP Based Model for Monitoring and Controlling Physical Parameters. 105
  Vishnu Kanthan Rathina Raj and Meena Belwal

Hardware Trojan Detection Using XGBoost Algorithm for IoT with Data
Augmentation Using CTGAN and SMOTE 116
  C. G. Prahalad Srinivas, S. Balachander,
  Yogesh Chandra Singh Samant, B. Varshin Hariharan,
  and M. Nirmala Devi

Data Analytics and Cloud Computing

Predictive Modeling of the Spread of COVID-19: The Case of India....... 131
  Sriram Sankaran, Vamshi Sunku Mohan, Mukund Seshadrinath,
  Krushna Chandra Gouda, Himesh Shivappa,
  and Krishnashree Achuthan

ASIF: An Internal Representation Suitable for Program Transformation
and Parallel Conversion..................................... 150
  Sesha Kalyur and G. S. Nagaraja

Performance Comparison of VM Allocation and Selection Policies
in an Integrated Fog-Cloud Environment 169
  M. R. Shinu and M. Supriya

**Artificial Neural Network, Machine Learning
and Emerging Applications**

Fraud Detection in Credit Card Transaction Using ANN and SVM 187
  Anchana Shaji, Sumitra Binu, Akhil M. Nair, and Jossy George

Detection of Leukemia Using K-Means Clustering and Machine Learning ... 198
  V. Lakshmi Thanmayi A, Sunku Dharahas Reddy, and Sreeja Kochuvila

An Analysis and Implementation of a Deep Learning Model
for Image Steganography..................................... 210
  Raksha Ramakotti and Surekha Paneerselvam

Abstractive Text Summarization on Templatized Data 225
  C. Jyothi and M. Supriya

Author Index 241

5G Networks, Millimeter Wave Communication Systems and Emerging Applications

Analysis of Generic Routing Encapsulation (GRE) over IP Security (IPSec) VPN Tunneling in IPv6 Network

Md. Raihan Uddin[1]([✉]), Nawshad Ahmad Evan[1], Md Raiyan Alam[2],
and Md. Taslim Arefin[1]

[1] Daffodil International University, Dhaka, Bangladesh
arefin@diu.edu.bd
[2] Texas A&M University - Kingsville, Kingsville, Texas, USA
md_raiyan.alam@student.tamuk.edu

Abstract. The virtual private network has become an essential technique used for providing a secure remote connection to exchange information over the Internet Protocol network. As the Internet Protocol version 4 and version 6 have different features and structures, the existing virtual private networks are modified to run in the new environment. This paper 'Simulation of Generic Routing Encapsulation (GRE) over IPsec VPN Tunneling on IPv6 Network' is simulated by using a GNS3 network simulator. Here, the solution provided for GRE over IPSec VPN between two remote offices equipped with IPv4 network where the WAN is connected with IPv6 network. In general, IPv6 does not support any kinds of IPSec VPN but the need for this VPN is high for encrypted data. So, this paper will demonstrate an examined method of using the GRE over IPSec VPN through the IPv6 network.

Keywords: GRE · IPv6 · IPSec VPN tunneling · IPv4

1 Introduction

IPV6 is ready to use for future technology that rapidly increases its deployment in the network sector. Features like auto configuration, a simplified header, Faster routing, Reduced network complexity, and many more that make it an easy pick for network administrators to choose ipv6 over ipv4 [1]. But the biggest drawback of IPv6 is not backward compatible. It lacks backward compatibility with the existing internet protocol, which is IPv4. The IPv6 network deployment is not done more in the network world. We need new transition tools until all network devices are compatible with the coexistence of IPv4 and IPv6.

A virtual private network is a technology for creating reachability between different private networks by establishing an end to end connectivity. And there are various tunneling protocols like Point to point tunneling protocol, Internet

© ICST Institute for Computer Sciences, Social Informatics and Telecommunications Engineering 2021
Published by Springer Nature Switzerland AG 2021. All Rights Reserved
N. Kumar et al. (Eds.): UBICNET 2021, LNICST 383, pp. 3–15, 2021.
https://doi.org/10.1007/978-3-030-79276-3_1

protocol security, Layer 2 tunneling protocol, Generic routing encapsulation, Secure socket layer to provide security for VPN tunnel. These protocols are working fine on the current IPV4 networks. But as mentioned earlier, the network world is moving forward rapidly from IPv4 to IPv6, where IPV6 promises better security and more advanced tunneling techniques have to be developed [2]. IPsec (Internet Protocol Security) is a protocol that provides a virtual private network service between sites. IPsec had designed to support a secure TCP/IP connectivity over the IP network by maintaining flexibility and scalability. IPsec is generally used for the service of cryptographic security. Due to maintaining data security through the VPN, the demand for IPsec is increasing [3]. IPsec proves the importance with various security features aiming at a secured data transmission between end devices. IPsec provides confidentiality of VPN traffic by encrypting it. It has also an encapsulation method that is known as Hashing algorithms. IPsec provides authentication services by using Digital certificates or Pre-shared keys. It protects against Reply attacks using a sequence number that built in the IPsec packets. By using these sequence numbers, IPsec can identify the packets which it already sent. IPsec can provide network security by encapsulating and encrypting traffic that passes through VPN from source to destination [4].

The backward compatibility issues of IPv6, this paper will analyze two of the existing VPN tunneling protocols, IPsec and GRE. The aim of this task to develop a standard and secure tunneling technique. That envisioned is a scenario where the remote end devices compatible with the existing Internet Protocol. Which can be connected with the internet and that incorporates IPV6 using GNS3 simulator. The following section on related works where we will discuss about the related publications and similar works on this topic.

As far as we know, no one has published any work about the implementation in Ipv6 network. Therefore, the research in this paper is kind of an innovative work.

2   Related Works

The transition from IPv4 to IPv6 is an active area of research. In [5] Various transition techniques like dual-stack, tunneling, and translation are compared based on quality parameters such as Average round trip time(RTT), bandwidth, and throughput.

Various IPv6 transition methods based on tunneling had an analytical discussion and later they were compared in [6]. A variety of parameters such as deployment time, CPE change, IPv4 continuity, access network, address mapping, end-to-end transparency, scalability was put into consideration in this paper.

4to6 and 6to4 are two widely used transition Mechanisms over Point to Point and IPsec VPN. Their performance was compared in different environments [7]. They have compared IPv4 or IPv6 with multiple transition mechanisms in terms of various parameters containing UDP and TCP traffic throughput, delay of the packets, jitter in the system, DNS, and VoIP with and without VPN.

The task in [8] proposes a new ISP Independent Architecture (IIA) for inter-connectivity in a hybrid network running with IPv4 and IPv6 networks. More flexibility issues of the users to deploy their required transition solutions are provided for their promised ISP. In their proposed model a network administrator can create multiple combinations of transition mechanisms for different destinations to manage security and load balancing.

Both in [9] and [10] the problems related to the transition between IPv4 and IPv6 are discussed. Problems like Interception by RA/DHCP server, Interception by Firewall. Unstable functions in the DNS zone record, poor coordinated tunnel network, lack of security in the path or peering, bad TCP connection, error in DNS resolution are taken into consideration in [9] causing issues like delay and disconnection. In contrast, the following section of our proposed system will represent the GRE over IPSec-VPN Tunneling over IPv6 using GNS3 where the mentioned paper has been shown the simulation of GRE over IPSec in IPv6.

3   Methodology

GRE over IPSec tunneling is a most used concept where this paper proposed in Fig. 1; a new method for the most advanced IPSec tunneling on the IPv6 network system. The way the procedure is elaborately explained below.

1. We will take two routers as source and destination and both will be CISCO routers.
2. Both source and destination will be connected via cloud or service provider and both routers loopback is reachable via the cloud.
3. GRE (Generic Routing Encapsulation) tunnel is based on the IPv6 loopback of the routers but we will use the IPv4 tunnel address as the peering address of the tunnel.
4. Then we will do IPv4 routing over IPv6 using the GRE tunnel and created another IPv4 loopback for both routers. Also the reachable IPv4 loopbacks by routing through the tunnel.
5. After that, we will configure GRE over IPSEC using that IPv4 loopback of our source and destination router.
6. We will set up our workstations (PC/Server) that will be configured by IPv4 address and check by ping and found that we can reach our destination successfully.
7. We will measure in Solar Winds that for ICMP traffic we are getting a bandwidth graph.
8. By Wire Shark we will check the packets are sending from source to destination are encrypted by ESP protocol. After finishing all measurements, we will be able to find our LANs are communicating and the IPv4 data get encrypted by GRE over IPSec which were passed through the service providers IPv6 network.

IPv4 and IPv6 are playing a big rule in the transition to create a platform where both ipv4 and IPv6 can co-exist. 3 basic transition mechanisms are mentioned below.

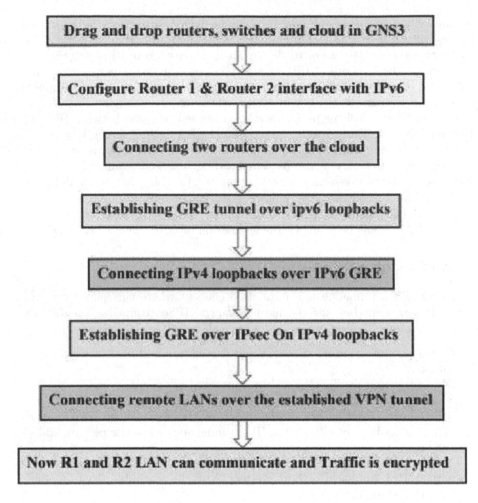

Fig. 1. Proposed working block diagram.

1. Dual Stack
2. Tunneling
3. Translation Techniques

1. ***Dual Stack.*** Dual-stack technology runs both IPv4 and IPv6 dual internet protocol in a single environment and it provides compatibility in communication for both routers and hosts [11]. This mechanism is put into action when the router interfaces are cable to process both IPv4 and IPv6 traffic. The interfaces are usually preferred to configure both IPv6 and IPv4 alongside. In Fig. 2, The Domain Name Server plays an important role to translate from Internet Protocol addresses to the domain name. For the dual-stack scenario, the IPv4 traffic communicates with the version 4 DNS server, and IPv6 communicates with the version 6 DNS server. To operate this mechanism, the

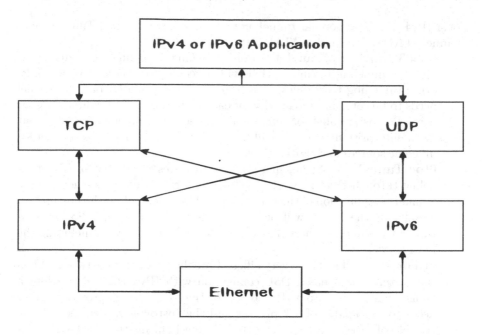

Fig. 2. Dual stack mechanism.

routers must be enabled in both IPv4 and IPv6 routing [12].

All recent IoT devices and operating systems, are now supported by both IPv4 and IPv6.

2. **_Tunneling._** Tunneling is a technique to create a point to point network connectivity from one network to another remote network for transferring data. For conveying IPvN packets through the IPvM network we configured a tunnel by measuring IPvN (Host1) as source and IPvN (Host2) as the destination. In Fig. 3, When the packets from IPvN travels via the tunnel it gets encapsulated by the tunnel header. Then the IPvN traffic is forwarded through the IPvM network. When the egress endpoint of the tunnel receives the encapsulated IPvN packet, it opens the encapsulated packet, extracts the original IPvN packet, and forwards it to the destination Host2 network. The tunnel header is set up before the original IP header. Generic Routing Encapsulation is the best tunneling technique that allows routing of IPv4 over IPv6 or IPv6

Fig. 3. Tunneling mechanism.

over IPv4 [12]. Five popular tunneling techniques are - Manual Tunnels, 6to4 tunnels, 6RD, GRE, and ISATAP [5].

- **6in4 Tunnel or Manual Tunnel** – The 6in4 tunneling mechanism provides a tunneling technique that will be configured manually [13]. Like other tunneling techniques, it will also be encapsulated but the tunnel needs to be configured manually for each connection. Scalability is absent here. This is suitable for individual connectivity for home or SME but in an enterprise network, it will be tough to maintain a lot of tunnel for different sources and destinations.
- **6to4 Tunnel** – 6to4 tunneling protocol provides scalability for IPv6 tunneling through IPv4 network [14] Like 6in4 this tunnel does not require manual configuration of the tunnel for each source. Only if the destination is different the tunnel will be different here. The encapsulation process will be the same as others that the IPv6 traffic will forward including the IPv4 header.
- **6rd or 6RD** – The 6rd is very efficient for the production network. When the customer end has a DSL connection with IPv4 the 6rd provides a translation process from IPv4 to IPv6. Here the service provider router needs to be capable of 6rd process and the customer router needs to be capable of IPv6 configuration. By this mechanism, the IPv4 addresses will convert into the hexadecimal format and set up a new IPv6 address. Then it travels via a tunnel which is configured in the service providers' router [15].
- **ISATAP** – ISATAP or Intra Site Automatic Tunneling Address Protocol. It forwards IPv6 packets over the IPv4 network using dual stack technology. ISATAP circuit set up a header for the IPv6 packet and passes the traffic via circuit or end to end tunnel [16].
- **GRE Tunnel** – The GRE Tunnel maintains its ideal technique to process tunneling mechanism and transfer traffic over the tunnel. GRE provides a point to point tunnel which includes 4 bytes GRE header before the original IP packet and by encapsulating the packet transfer to its destination. This is one of the most used VPN for routing IPv4 over IPv6 network and vice versa [17].

3. *Translation Techniques.* Direct communication between IPv4 and IPv6 is achieved by the IPv4-IPv6 translation technique. In Fig. 4, The basic principle of translation is shown in the figure below. The idea is to convert the semantics between Here er can measure IPvM as IPv4 and IPvN as IPv6. Generally, translation happens when an IPv4 device wants to communicate with an IPv6 destination. For IPv6 translation, the IPv4 is converted to a hexadecimal state and set up in IPv6 format maintaining the last 64 bits and first 32 bits. The response will return to IPvM like vice versa. Again if the IPv6 device wants to communicate with the IPv4 device an additional IPv4 header will add to encapsulate the IPv6 packet. The NAT router will operate this process of translation. The domain name service will also do the translation by exchanging information with each other. For this one translator server will

be added before IPv6 and IPv4 DNS server [18].

Some of the very popular translation technique are discussed below:

- **NAT-PT** IPv4 and IPv6 bidirectional communication happened via translation and NAT-PT translator has the exact feature for this. This also works in protocol level and translates Internet Protocol, Domain Name Server, and Internet Control Message Protocol [19].
- **NAT64 and DNS64** Network Address Translation 64 DNS64 [20] is a popular translation technique for IPv4 and IPv6. It enables IPv6 hosts to communicate with IPv4 servers or workstations and IPv4 to IPv6. The received IPv4 packets are converted by the NAT64 translator by converting IPv4 hexadecimal bits in IPv6 Hextet for developing a new IPv6 address to communicate with IPv6 servers. IPv6 DNS also has a NAT64 DNS server that allows communication with both IPv4 DNS and IPv6 DNS.
- **464 XLAT** For both stateful and stateless translation 464XLET is used in the enterprise network. For translating IPv4 to IPv6 this technique has the advance feature and scalability more than NAT64 and DNS64 due to its fast deployment. The major thing is that to run this no new protocol is needed in the enterprise network [21].

Fig. 4. Translator mechanism.

4 Implementation

This paper has considered a scenario where two remote branches, one located at Dhanmondi and another located at Ashulia having network equipment compatible with IPv4, are connected via the internet of the IPv6 network shown in Fig. 5. The network simulation is done using the GNS3 version: 2.2.5 consisting of routers with the Cisco 7200 VXR series and switches with Catalyst 2600 series. Now at the Dhanmondi office, we are considering two loopback addresses of router 1. Where loopback address 10.1.1.100 is configured using IPv4 and loopback address 4004:1/128 is configured using IPv6.

At the Ashulia office, we have also considered two loopback addresses for router 2. Loopback address 10.1.1.200 is configured using IPv4 while the other loopback address 4004.1/128 is configured using IPv6. After connecting the routers over the cloud we have established a GRE tunnel over IPv6 loopbacks of both routers. Then we have connected IPv4 loopbacks over IPv6 GRE tunnel establishing GRE over IPsec on IPv4 loopbacks. Thus two remote LANs are connected over the established VPN tunnel. Now router 1 LAN and router 2 LAN are communicating and transferring data. The data is getting encrypted by IPSec parameters and for GRE parameter we can set any static route over the tunnel. The IPv6 has a drawback that it does not support the IPSec tunnel for that reason we make a hybrid solution for that. We did IPv4 communication over the IPv6 service provider network and used GRE over IPSec VPN to encrypt our data and easy routing process for communication of IPv4 workstations at the source and destination premises. All measurement parameters and their results are given here.

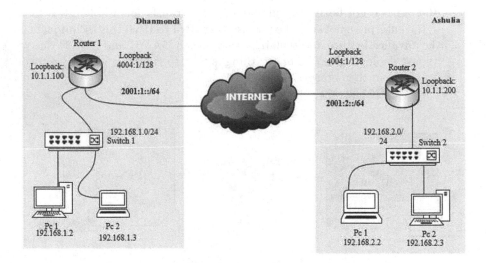

Fig. 5. Implemented network topology.

4.1   Graphical Networking Simulator (GNS3)

Network equipment can be easily installed and available in Graphical Network Simulator or GNS which also allows critical network emulation. When it is installed with VMware or Virtual Box we can collected and run the different operating systems in a virtual environment and can get the test of real infrastructure. This program allows different systems and devices to communicate with each other and a user can create topology according to his planning. Cisco devices can be deployed as Dynamaps in GNS and by using command lines we can operate it. Dynamics are the program of CISCO IOS which is designed for running in a virtual environment. GNS3 provides a graphical environment to run different networks and systems.

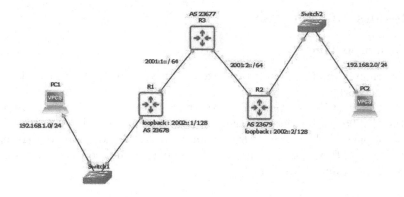

Fig. 6. Network topology in GNS3 simulator.

4.1.1   WireShark
Wireshark is an important tool that is used for analyzing packets. By using Wireshark, we can capture the network packets and analyze the detailed information contained in the packets. This is also used for deep troubleshooting. We can measure source information, destination information, protocol, packet type, and other necessary information graphically.

4.1.2   Solar Putty
Solar putty provides us the command line interface with which we can operate the devices. To configure any device, we may need a Console and an emulator terminal to generate command. Solar Putty provides us an open source terminal emulator and Console for the serial port as well as we can use different services from that like SSH, Telnet, SCP, and socket connection.

4.1.3   Solar Winds
We can measure the network bandwidth and network performance by Solar Winds. In this project, we captured the network bandwidth by sending ICMP traffic from source to destination. Solar Winds is the tool for monitoring traffic bandwidth and analyze in real time.

4.2   Design and Analysis in GNS3

In Fig. 6, We designed a model in the GNS3 network simulator and implemented a real model for our planned network topology. We used both IPv4 and IPv6, Cisco routers, Cisco Switches, and PC as end devices in the virtual environment. The diagram is displayed here.

5   Results and Analysis

5.1   Response Time

Response time is the travel time of packets from one source to destination in a computer network. In Fig. 7 We will calculate the response time from the Packet Internet Gopher (PING) report. The lowest response time is 35.795 ms and the highest is 47.795 ms. So, the average for five response time is 40.372 ms. Here the type of packet which travels from source to destination is ICMP (Internet Control Message Protocol). We measured, calculated, and collected the data by sending ICMP traffic in Table 1.

5.2   Throughput

The highest production or extreme production rate is measured as throughput at which we can produce something. In Ethernet technology or computer networks, we can measure throughput with different parameters. Successful transmission of any packet depends on both physical and logical connectivity. In Fig. 8, Generally, we calculate the data transfer via an interface in bps, Kbps, Mbps, and Gbps. Here we found a throughput in Kbps after transferring ICMP traffic from source for router-1 (source router) out interface.

```
PC1> ping 192.168.2.2
84 bytes from 192.168.2.2 icmp_seq=1 ttl=62 time=36.789 ms
84 bytes from 192.168.2.2 icmp_seq=2 ttl=62 time=38.298 ms
84 bytes from 192.168.2.2 icmp_seq=3 ttl=62 time=35.795 ms
84 bytes from 192.168.2.2 icmp_seq=4 ttl=62 time=43.254 ms
84 bytes from 192.168.2.2 icmp_seq=5 ttl=62 time=47.725 ms

PC1> show ip

NAME          : PC1[1]
IP/MASK       : 192.168.1.2/24
GATEWAY       : 192.168.1.1
DNS           :
MAC           : 00:50:79:66:68:00
LPORT         : 20024
RHOST:PORT    : 127.0.0.1:20025
MTU:          : 1500

PC1> []
```

Fig. 7. Ping report.

Table 1. Responses of the packet sequences

Packet type	Sq.	Source (host)	Dest. (host)	Response time (ms)	Avg. response time (ms)
ICMP	1	192.168.1.2	192.168.2.2	36.789	40.372
	2			38.298	
	3			35.795	
	4			43.254	
	5			47.725	

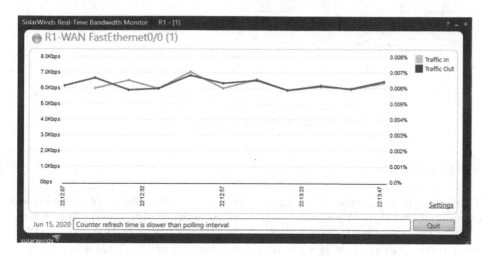

Fig. 8. Throughput analysis.

5.3 Packet Analysing

We can analyze our network traffic by Wireshark and can analyze packets that we are sending. As we are using IPSec, the ESP protocol will encrypt our traffic and it will be completely unable to analyze data from encrypted packets showed in Fig. 9. Hereafter, the crypto session established we analyze the encrypted packets via Wire Shark.

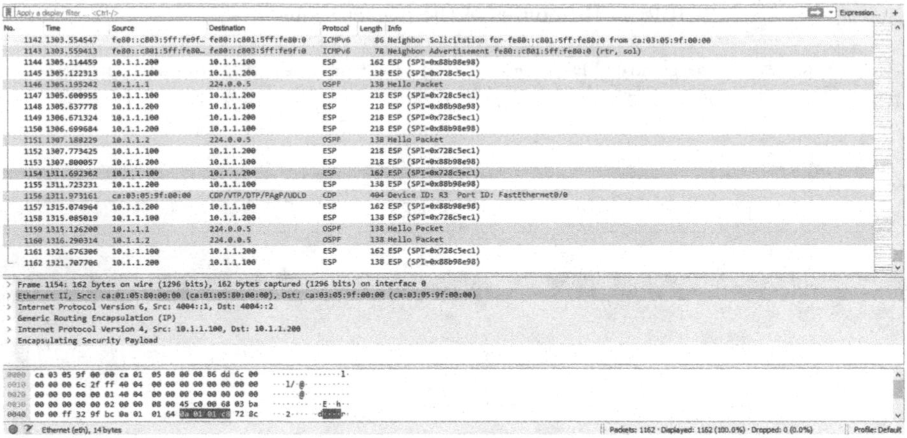

Fig. 9. Packet analyzing.

6 Conclusions

This paper aims to connect to remote offices that were using IPv4 compatible equipment and their WAN network is an IPv6 network. To combine the IPSec VPN tunneling technology and IPv6 we simulated GRE over IPSec as a VPN tunnel using GNS3. This is a prominent solution to the fact that most of the network equipment is still IPv4 compatible while the world is rapidly moving towards the deployment of IPv6 worldwide. The translation between IPv4 and IPv6 is very important as we can not ignore the IPv4 when we had to move forward with IPv6. Again IPv6 does not support the most popular VPN for encryption. This paper demonstrates a clear solution for encrypted data transfer of IPv4 over a new generation Internet Protocol.

References

1. Ahmad, N.M., Yaacob, A.H.: IPSec over heterogeneous IPv4 and IPv6 networks: issues and implementation. Int. J. Comput. Netw. Commun. **4**(5), 57 (2012)
2. Venkateswaran, R.: Virtual private networks. IEEE Potentials **20**(1), 11–15 (2001)
3. Tjahjono, D., Shaikh, R., Ren, W.: U.S. Patent and Trademark Office. U.S. Patent No. 8,893,262. Washington, DC (2014)
4. Yildirim, T., Radcliffe, P.J.: VoIP traffic classification in IPSec tunnels. In: 2010 International Conference on Electronics and Information Engineering, vol. 1, pp. V1–151. IEEE, August 2010
5. Singalar, S., Banakar, R.M.: Performance analysis of IPv4 to IPv6 transition mechanisms. In: 2018 Fourth International Conference on Computing Communication Control and Automation (ICCUBEA), pp. 1–6. IEEE, August 2018
6. Kim, P.S.: Analysis and comparison of tunneling based IPv6 transition mechanisms. Int. J. Appl. Eng. Res. **12**(6), 894–897 (2017)

7. Narayan, S., Ishrar, S., Kumar, A., Gupta, R., Khan, Z.: Performance analysis of 4to6 and 6to4 transition mechanisms over point to point and IPSec VPN protocols. In: 2016 Thirteenth International Conference on Wireless and Optical Communications Networks (WOCN), pp. 1–7. IEEE, July 2016
8. Saraj, T., Yousaf, M., Akbar, S., Qayyum, A., Tufail, M.: ISP independent architecture (IIA) for IPv6 packet traversing and inter-connectivity over hybrid (IPv4/IPv6) internet. Procedia Comput. Sci. **32**, 973–978 (2014)
9. Hirorai, R., Yoshifuji, H.: Problems on IPv4-IPv6 network transition. In: International Symposium on Applications and the Internet Workshops (SAINTW 2006), pp. 5-pp. IEEE, January 2006
10. Ahmed, A.S., Hassan, R., Othman, N.E.: Security threats for IPv6 transition strategies: a review. In: 2014 4th International Conference on Engineering Technology and Technopreneuship (ICE2T), pp. 83–88. IEEE, August 2014
11. Haider, A., Houseini, M.: The difference impact on QoS parameters between the IPSec and L2TP. Int. J. Innovative Adv. Eng. (IJIRAE) **11**(3), 31–42 (2016)
12. Wu, P., Cui, Y., Wu, J., Liu, J., Metz, C.: Transition from IPv4 to IPv6: a state-of-the-art survey. IEEE Commun. Surv. Tutorials **15**(3), 1407–1424 (2012)
13. Babatunde, O., Al-Debagy, O.: A comparative review of internet protocol version 4 (IPv4) and internet protocol version 6 (IPv6). arXiv preprint arXiv:1407.2717 (2014)
14. Huijun, D.U.: Application of 6to4 tunnel technique based on dual stack. J. Guangdong Polytechnic Normal Univ. **12** (2007)
15. Yoon, S.J., Park, J.T., Choi, D.I., Kahng, H.K.: Performance comparison of 6to4, 6RD, and ISATAP tunnelling methods on real test beds. Int. J. Internet Distrib. Comput. Syst. **2**(2) (2012)
16. Guo, L.L., Guo, Y.M., Wang, Y., Dong, N.: Implementation of IPv6 network based on combination of 6to4 and ISATAP tunnel techniques. Radio Commun. Technol. **3** (2006)
17. Sansa-Otim, J.S., Mile, A.: IPv4 to IPv6 transition strategies for enterprise networks in developing countries. In: Jonas, K., Rai, I.A., Tchuente, M. (eds.) AFRICOMM 2012. LNICST, vol. 119, pp. 94–104. Springer, Heidelberg (2013). https://doi.org/10.1007/978-3-642-41178-6_10
18. Gilligan, R.E., Nordmark, E.: Basic transition mechanisms for IPv6 hosts and routers. In: IETF RFC (2005)
19. Aravind, S., Padmavathi, G.: Migration to Ipv6 from IPV4 by dual stack and tunneling techniques. In: 2015 International Conference on Smart Technologies and Management for Computing, Communication, Controls, Energy and Materials (ICSTM), pp. 107–111. IEEE, May 2015
20. Raicu, I., Zeadally, S.: Evaluating IPv4 to IPv6 transition mechanisms. In: 10th International Conference on Telecommunications, ICT 2003, vol. 2, pp. 1091–1098. IEEE, February 2003
21. Al-Azzawi, A., Lencse, G.: Towards the identification of the possible security issues of the 464XLAT IPv6 transition technology. In: 2020 43rd International Conference on Telecommunications and Signal Processing (TSP), pp. 439–444. IEEE, July 2020

Mobility Management Based Mode Selection for the Next Generation Network

Pallavi Sapkale[1]([✉]), Uttam D. Kolekar[2], and Navin Kumar[3]

[1] Electronics and Telecommunication Department, RAIT Nerul, Navi Mumbai, India
pallavi.sapkale@rait.ac.in
[2] Electronics and Telecommunication Department, APSIT Thane, Thane, India
[3] Department of ECE, Amrita School of Engineering Bengaluru,
Amrita Vishwa Vidyapeetham, Bengaluru, India
navinkumar@ieee.org

Abstract. In this work, we proposed a modish approach to enhance Quality of Service (QoS) in the mobility management of cellular mobile communication using mode selection method. In this approach, mode selection method uses certain quality parameters such as delay, signal strength and throughput. Different mode selection scheme is proposed based on accessibility and decision making, i.e., either the User Equipment (UE) or the Cellular Mode (CM). During the exchange of the network from cellular mode to user mode, there is a need to maintain the acceptable quality of parameters. It is decided by mobility management model. The proposed method offers faster response (0.165 s lesser than the existing) and 1.64 Mbps throughput.

Keywords: Handover · LTE · Mobility management · Mode selection · Quality of Service (QoS) · Wireless network

1 Introduction

The demand of mobile data is increasing exponentially and in the fifth generation (5G) cellular mobile communication systems, the data speed is expected to reach up to 10 Gbps, around 20 fold increases from that of forth generation (4G) Long Term Evolution Advanced (LTE-A) [1]. The speedy growth in the usage of mobile communication has greatly necessitated the expansion of cellular mobile. At the same time, it is expected that the network to provide good Quality of Service (QoS). From Ericsson mobility report (June 2020), the average traffic for each smartphone is anticipated to increase nearly 25 Gb per month in 2025 [2]. Furthermore, near about 410 million smartphone users are anticipated in India by 2025, wherein global traffic rate would reach around more than 35 Eb per month by the end of 2020 [2]. More mobility support devices are used continuously by users. One of the Quality of Service (QoS) parameter is seamless

N. Kumar et al. (Eds.): UBICNET 2021, LNICST 383, pp. 16–25, 2021.
https://doi.org/10.1007/978-3-030-79276-3_2

connectivity, that is, mobility which can be addressed by mobility management [3]. Because of denser deployment, users require much frequent handover. The handover can be performed by User Equipment (UE) also called user mode or by base station, also called cellular mode. In addition, it is also possible to perform the function sometimes by cellular mode while sometimes, with user mode. Mobility management plays important role in the mode selection. However, frequent and inefficient process of mode selection between user mode to cellular mode especially when handover process initiates, degrades the network performance. UE system does not have the intelligence to select the best mode to enhance or maintain the Quality of Service (QoS).

In this paper, we present an enhanced QoS using mobility management to assist selection of modes like user or cellular for seamless connectivity. The switching of modes is enabled on the basis of Quality of Service (QoS) parameter like signal strength and delay of user mode or cellular mode. Enhance QoS Mobility Management (eQMM) is checked for heterogeneous networks. It selects the calls from user mode to cellular mode. It is based on network performance. During the selection of mode, mobility management will insist to choose for better performance so that QoS is maintained. And, according to the selection process, call will be transferred towards the best network. Our proposed work build up on customers QoS performance such as less communication delay and maximum throughput. We scrutinize these terms and formulate them with switching factors.

Rest of the contents of the paper is arranged as follows. In Sect. 2, mobility management in cellular network and, mobility management model is discussed, propose mode selection method is explained in Sect. 3. Section 4 presents results and discussion while conclusion is given in Sect. 5.

2   Mobility Management in Cellular Network

2.1   Relevant and Recent Work

In this section, we describe the overview of present state of art for mobility management in cellular network and consider LTE-A as recent technology.

In Fig. 1, basic cellular network is shown, where base station includes antenna, controller and number of receivers. Base Transceiver Station (BTS) is installed on upper side of the antenna. It is a mobile phone access point to the network. BTS communicate between network and mobile phone. While Mobile Telecommunication Switching Office (MSTO) connects calls between mobile units. It help to set up the calls and handle the call routing and call switching also. MSTO is also helpful for the handover. Mainly two types of channels are available between mobile unit and base station.

1. Control Channel: This is used for exchanging and maintaining the information.
2. Traffic Channel: This is use to carry the voice as well as data connection in between the clients or users.

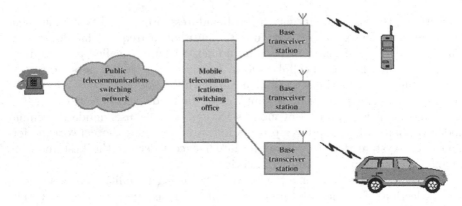

Fig. 1. Basic cellular network

In the past years many research works addressed the challenging issue of mobility management in cellular network to improve the Quality of Service (QoS) with respect to parameters such as delay, jitter, interactive traffic and low Bit Error Rate (BER). Authors in [4], designed the mobility prediction coupled with the various protocol which was used to improve QoS for mobile devices and explained the traffic load distribution in Wi-Fi network without disturbing. In [5], authors developed an enhanced mobility management and Vertical Handover (VHO) algorithm which was used during Device to Device (D2D) communication for maximizing the throughput. However, the scheme is for precise borders of different regions with less time. Authors in [6] addressed a device centric communication for networks. [12] designed Location Management (LM) system for the personal communication system. This method proposed for mode writing in device to the network. But in this work, movement of user in mobile management is not investigated and discussed. In some research [6], user centric mesh is used to determine the dynamic femto access node group for 5G mesh but again node mobility is not addressed. In this paper, we present an enhancing Quality of Service (QoS) mobility management using mode selection algorithm.

2.2 Mobility Management Model and Architecture

Figure 2 shows the mobility management architecture. This model includes two parts. One is handover unit for controlling the handover and Context Aware Module (CAM) for generating trigger event for handover. Handover unit examines the better selection of network and grant the decision towards Cellular Mode (CM) or User Mode (UM) and stores the remaining information in data storage unit. Figure 3 shows handover stages in mobility management module. Handover procedure depends on three major actions,

1. One is collecting the information in which it checks the availability for network and also find the user requirement.

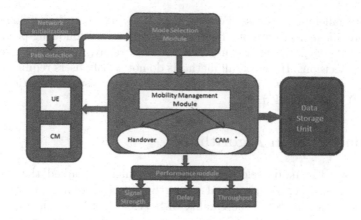

Fig. 2. Proposed mobility management model using mode selection method

2. Decision takes place using resources of availability and from the decision algorithm handover decision takes place. In final stage handover execution occurs.

3   Proposed Mode Selection Method

This is a mode selection method where mesh changes towards user centric transmission. This method is independent on mesh infrastructure wherein system to system communication is not depended on enhance Node Base Station (eNB). The proposed work has outcome based mode selection using mobility management. Figure 3 shows proposed mode selection method.

Initially, the network is established and searches a path for further communication. Once the path is detected, the path detection performs the input function for Mode Selection Module (MSM). In wireless networks, the selection of best

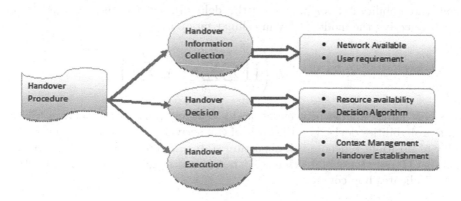

Fig. 3. Handover stages in mobility management module

network is challenging process. For seamless mobility there is no single factor that can provide a clear data of which term to select. The following performance parameters like signal strength, user satisfaction, delay, maximum throughput are used for decisions. However, all methods do not satisfy users requirements. In this current proposal, by minimizing delay and latency user satisfaction Quality of Service (QoS) is fulfilled.

4   Mode Selection Algorithm

Figure 4 shows the mode selection algorithm which explained the process of selection. Let $s(g)$ represents signal strength and $b(w)$ represents bandwidth. These operations will record the Quality of Service (QoS) specifications such as UE and Correspondent Node (CN) to calculate the signal strength between UE to eNB and Correspondent Node (CN) to eNB. And the network for UE and CN is denoted by $N(w)$. Also, consider performance of energy for the signal is denoted as $e(n)$ and time delay to reach eNB is denoted as t(d). If signal is maximum from threshold value (Δ) then call will move towards UM. We used threshold values same as LTE network. P_r is indicates for the performance parameters. And the e(n) used for energy while t(d) is denotes for time delay.

4.1   Model Formulation with Mobility Management

Consider client motion probability is MB, a cell phone client walk in limited service zone D. Starting position of the client is assumed as $U_1(a_1b_1c_1)$ and actual location is $U_2(a_2b_2c_2)$, the $distance = |U1U2| = \sqrt{(a_2 - a_1)^2(b_2 - b_1)^2(c_2 - c_1)^2}$. For each movement of the user will change the results.

4.2   Delay

In the proposed method, the eNB will select the mode of communication on the basis of network performance. So $delay = Time1 + Time2$. Due to free user movement, the mobility and hopping affect this delay. In between the communication user is selecting the mode. Delay in cellular mode is,

$$t(d)_{CM} = Ug_e \left(\prod_k \sum_{e=1}^{i} \sum_{j=0}^{\gamma} \frac{\partial J}{\partial J} \rho + \pi_i^k \sum_{j=0}^{\frac{\partial J}{\partial J}} \rho \right) \tag{1}$$

Where, U indicate movement factor, and ρ denotes transmission delay. g_e represents probabilities devised on the basis of call arrival and departure, i represents index of position, k denote call departure states, J represents per hop link utilization, $\prod$ represents state change, j represents signifies movement direction and γ indicates hop count.

Mode Selection Algorithm

Data: s(g) ⟶ UE, s(g)⟶CN, b(w)⟶UE, b(w) ⟶ CN

Device mode selection

Calculate: P(r)[UE, eNB], P(r)[CN, eNB]

While P(r) [UE, eNB], P(r)[CE, eNB] = Threshold(?)

User -Mode();

End

Cellular- Mode();

Data N(w)⟶ UE , N(w)⟶ CN

Performance $P_{r(a)}$

Calculate e(n) ⟶ [UE, eNB], t(d) [UE, eNB]

t(d) [CN, eNB], b(w) (β)

If e(n) ⟶[UE, eNB] && e(n) [CN, eNB]

&& t(d) [UE, eNB] &&

t(d) [CN, eNB] && b(w) = Threshold(?)

User Mode();

Data: D(S) for UE, CN

If Ds = Threshold(?)

Calculate N(w) ⟶UE, N(w) ⟶ CN

The Enb calculates

S(g),t(d),e(n).

P(R)[v] == S(g) && t(d) && e(n) = Threshold(?) User Mode (UM)();

Otherwise Cellular mode(UM)

Γ— Metric between source node to target node.

Fig. 4. Algorithm for mode selection

This delay in UM is expressed by,

$$t(d)_{UM} = U g_e \left(\prod_k \sum_{e=1}^{i} \sum_{j=0}^{\psi} \frac{\partial J}{\partial} \rho + \pi_i^k \sum_{j=0}^{\frac{\partial J}{\partial}} \rho \right) \qquad (2)$$

4.3   Energy

The usage of energy is based on network energy. Which is dependent on mobility factor. Energy usage in CM describe as,

$$e(n)_{CM} = \prod_k \sum_{e=1}^{\gamma} \sum_{h=1}^{U} (\alpha + \beta + x) + \prod_k \sum_{e=1}^{\gamma} \sum_{h=1}^{U} x \tag{3}$$

where, α denotes group energy, γ represents hopping, U denotes signifies user mobility factor such that $1 \leq h \leq U$, β indicates network update energy and x is signaling cost. The energy usage in UM is expressed as,

$$e(n)_{UM} = \prod_k \sum_{e=1}^{\gamma} \sum_{h=1}^{U} (\alpha + \beta + x) + \prod_k \sum_{h=1}^{U} x \tag{4}$$

5   Simulation Setup

The performance result of the proposed method is done in MATLAB R18a using PC with Windows 10 Operating System, 2 GB RAM, and Intel i5 core processor. Parameters such as utilizing delay, power and throughput are evaluated. We used 110 nodes and simulation time is 500 s. Following Fig. 5 shows output graphs for the delay and Fig. 6 shows comparison for throughput. Figure 7 shows output graphs for power.

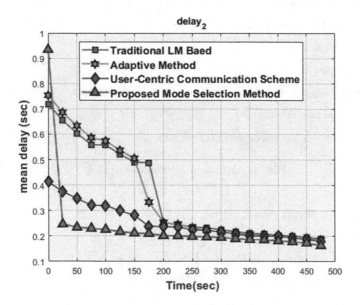

Fig. 5. Delay in proposed method

Table 1. Results

Metrics	Traditional location based method	Adaptive scheme	User-centric communication method	Proposed mode selection method
Delay (s)	0.195	0.188	0.180	0.165
Power	28.448	33.810	36.768	58.785
Throughput (bps)	1138543	1550007	836393	1641723.5

6   Result

The methods used for the comparative analysis are traditional Location Management (LM) method, adaptive method and user centric communication method. The comparison analysis for traditional method and proposed method with delay parameter is shown in Fig. 5. When simulation time is 470 s, the delay values calculated by the Location Management (LM) method, adaptive method, user centric communication scheme, and proposed mode switching methods are 0.195 s, 0.188 s, 0.180 s, 0.165 s. The analysis method for calculating throughput is shown in Fig. 6. When simulation time is 470 s, throughput values computed by traditional Location Management (LM) based is 1138543, adaptive method is 155007, user centric communication method is 836393 and in proposed mode selection method throughput is maximum value as 1641723. The analysis of methods using the power parameter is considered in Fig. 7, when simulation time is 470 s, the power values computed by the Location Management (LM) scheme, Adaptive scheme, User centric communication scheme, and proposed mode switching method are 28.448, 33.810, 36.768,58.785 (Table 1).

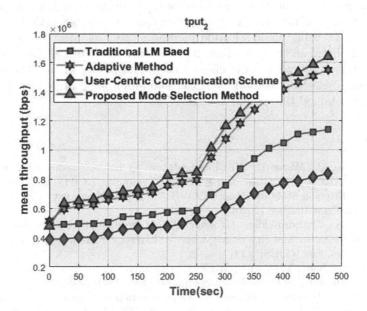

Fig. 6. Throughput in proposed method

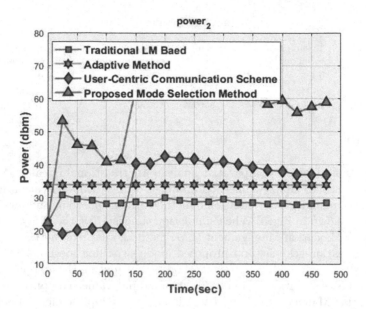

Fig. 7. Power in proposed method

7   Conclusion

A new mode selection scheme in cellular networks for mobility management is developed. At this point, in the system design the selecting factors, which includes secure execution based specifications, like delay, energy usage, and signal strength for available user. The enhancing Quality of Service (QoS) in mobility management using mode selection method as well as client mode selection also done. In between the procedure, it select user mode or cellular mode. For increasing the quality of service, client motion can be anywhere. The planned work of mode selection method gives the minimum delay of 0.165 s, power of 58.785, and obtained throughput is 1641723.5.

References

1. Li, Y., Cao, B., Wang, C.: Handover schemes in heterogeneous LTE networks: challenges and opportunities. J. IEEE Wirel. Commun. **23**, 112–117 (2016)
2. Ericsson: Ericsson Mobility Report: COVID-19 impact shows networks crucial role in society. Press Release (2020)
3. Reddy, D.S., Chandrasekhar: Generalized light gradient boost classifier for traffic aware seamless mobility management in heterogeneous network. J. Indian J. Comput. Sci. Eng. (IJCSE) **11**(1) (2020)
4. Yap, K.-L., Chong, Y.-W., Liu, W.: Enhanced handover mechanism using mobility prediction in wireless networks. PloS One **15**(1), e0227982 (2020)
5. Morattab, A., Dziong, Z., Sohraby, K.: Mode selection map-based vertical handover in D2D enabled 5G networks. J. IET Commun. **13**(14), 2173–2185 (2019). ISSN 1751-8628

6. Mustafa, H.A.U., Imran, M.A., Shakir, M.Z., Imran, A., Tafazolli, R.: Separation framework: an enabler for cooperative and D2D communication for future 5G network. Journal IEEE Commun. Surv. Tutor. **5**, 419–445 (2016)
7. Yang, B., Wang, X., Qian, Z.: A multi-armed bandit model based vertical handoff algorithm for heterogeneous wireless networks. J. IEEE Commun. Lett. **22**, 2116–2119 (2018)
8. Deswal, S., Singhrova, A.: A vertical handover algorithm in integrated macrocell femtocell networks. J. Int. J. Electr. Comput. Eng. (IJECE) **7**(1), 299–308 (2017)
9. Semiari, O., Saad, W., Bennis, M.: Caching meets millimeter wave communications for enhanced mobility management in 5G networks. J. Trans. Wirel. Commun. **17**(2), 779–793 (2018)
10. Sapkale, P., Kolekar, U.D.: Mobility management for 5G mobile networks. J. Int. J. Comput. Appl. **182**(26), 1–4 (2018)
11. Sapkale, P., Kolekar, U.: Handover decision algorithm for next generation. In: Vasudevan, H., Gajic, Z., Deshmukh, A.A. (eds.) Proceedings of International Conference on Wireless Communication. LNDECT, vol. 36, pp. 269–277. Springer, Singapore (2020). https://doi.org/10.1007/978-981-15-1002-1_28
12. Biswash, S.K., Sarkar, M., Sharma, D.K.: Artificial immune system (AIS)-based location management scheme in mobile cellular networks. Iran J. Comput. Sci. **1**(4), 227–236 (2018)
13. Isabel Sanchez, M., et al.: Mobility management: deployment and adaptability aspects through mobile data traffic analysis. Comput. Commun. **95**, 3–14 (2016)

Specification of a Framework, Fully Distributed, for the Management of All Types of Data and the Services Close to Users

Thierno Ahmadou Diallo[✉]

LI3, Assane SECK University of Ziguinchor, BP 523, Ziguinchor, Senegal
t.diallo@univ-zig.sn

Abstract. This article presents GRAPP&S (Grid APPlication & Services), a specification of a multi-scale architecture for the management (unified storage and indexing) of data and services near users. We manage all types of data and services through the use of specific node called proxy. GRAPP&S's architecture consists of three types of nodes, each with different roles. These nodes are grouped together in the form of communities (local networks) using multi-scale principles. The data is presented transparently to the user through proxies (an example of GRAPP&S nodes) specific to each type of data. In addition, the GRAPP&S architecture has been designed to allow the interconnection of different communities and the application of security and privacy policies, both within a community and between different communities. Our framework adopts a routing mechanism prefixed for research and access to data GRAPP&S. This access does not depend on a direct connection between the nodes, as in most P2P or other networks. In GRAPP&S, it is always possible to route the data transfer path used when looking at cases where a direct connection between the nodes is not possible.

Keywords: Hierarchical system · Proxies · Hierarchical addressing · Prefix routing

1 Introduction

Large-scale data management is a recurring problem in both science and business. Despite constant advances in memory and disk capacity, the use of a single storage device is no longer an option as competitive access, reliability, energy consumption and cost are obstacles to development systems. It is for this reason that researchers and developers have long turned to the development of distributed storage solutions, in order to work around these limitations.

In cases where the data can be represented under the In the form of files, NAS/SAN type solutions, P2P networks and also cloud storage represent technological choices capable of offering large-scale storage at a reasonable cost. These

N. Kumar et al. (Eds.): UBICNET 2021, LNICST 383, pp. 26–44, 2021.
https://doi.org/10.1007/978-3-030-79276-3_3

choices also apply to relational or NoSQL databases, but in this case access to the data always requires a (pseudo) centralized entity capable of aggregating and presenting the data (this does not exclude the parallel processing of data requests).

Through different strategies, these distributed solutions offer transparent solutions to increasing storage needs, while offering sufficient guarantees to ensure the consistency and sustainability of data. Today, the use of NAS/SAN or cloud file storage solutions has become as common as the use of USB drives or drives, once the potentially unlimited resources offered by P2P or cloud solutions present themselves several advantages in terms of cost, availability and use of physical resources. However, these solutions can also have drawbacks related to speed of access and data security; the solution to these drawbacks is still far from guaranteed and depends mainly on proprietary solutions offered by storage service providers. Another aspect to consider is the compatibility between the systems: if some APIs make the manipulation of files relatively simple, it is less obvious the integration of other data representations such as requests on a database, data streams or the performance of services. It is with the aim of providing a unified architecture for data and services that we present GRAPP&S (GRid Applications and Services), a multi-scale architecture for the aggregation of data and services. This framework has been designed to transparently integrate file type data as well as databases, streams (audio, video), Web services and distributed computing. Through a hierarchical structure based around the concept of "communities", GRAPP&S allows the integration of information sources with heterogeneous access protocols and various security rules.

Other contributions from GRAPP&S are the establishment of a separate specification of the communication middleware (P2P or otherwise), and the definition of a generic storage solution. In the first case, this is possible because the specification of GRAPP&S is based on properties such as management of community connectivity and management of routing between nodes. In the second case, the use of "proxy data" nodes makes it possible to unify access to various resources such as files, streams, services (FTP, mail) or data created on the fly by calculation spots or sensors.

Finally, data access is not dependent on a direct interconnection between nodes, as is the case with most P2P networks. In GRAPP&S, it is always possible to route the data transfer by the path used during the search, in case a direct connection between the nodes is not possible.

The remainder of this article is organized as follows: Sect. 2 briefly introduces the state of the art regarding multi-scale systems and P2P networks. Section 3 presents the main elements that make up the GRAPP&S architecture and how these elements interconnect, while Sect. 4 details the main operations related to managing and retrieving information in GRAPP&S. Section 5 details the current state of development of GRAPP&S and introduces the GAIA module for optimizing multi-scale storage [3, 14]. Finally, Sect. 6 offers our conclusions.

2 Releated Work

2.1 Multi-scale Systems

In general, multi-scale systems are therefore structured around services deployed on several levels and which are completely in order to meet the more or less immediate needs of customers. Rottenberg et a [1] formalize this definition on the form: A multiscale system is a system distributed over several levels of different sizes in one or more dimensions (equipment, network, geography, etc.). In [3] presents an example of a multi-scale system in his work around "cloudlets". In this work, the authors examine the limitations of mobile devices and the inadequacy of the solutions currently in place for the outsourcing of mobile services (in particular the transfer of these services to cloud computing). The proposed solution is the implementation of stateless proximity servers connected to the Internet (cloudlets) to which mobile devices can connect directly via a Wi-Fi network. These devices are deployed as Wi-Fi hotspots in cafes, shops, etc. They have high computing power and allow mobile devices to outsource overly complex calculations to a nearby machine, which solves cloud computing latency issues. This paper [10] has proposed an application-aware cloudlet selection strategy for multi-cloudlet scenario. Different cloudlets are able to process different types of applications. To reduce latency and work overload in mobile clouds, the authors propose, strategy can balance the load of the system by distributing the processes to be offloaded in various cloudlets.

2.2 Peer-to-Peer (P2P) Networks

In [5] this work, propose a novel P2P-based MCS architecture, where the sensing data is saved and processed in user devices locally and shared among users in a P2P manner. This article fixes, the traditional server-client MCS architecture often suffers from the high operational cost on the centralized server (e.g., for storing and processing massive data), hence the poor scalability. Gassara and al. [4] propose a resource management middleware in a multi-scale context. The aim of this system is to offer flexible management of a set of resources distributed in a multi-scale manner. In this article, the term multiscale is used to describe the distribution of resources across networks of different sizes: PAN (Personal Area Network), LAN (Local Area Network), WAN(Wide Area Network). The proposed solution consists of grouping the resources into domains and domain federations in order to implement different management models. The authors propose a formal approach supporting the correct description of deployment architectures and their reconfigurations. According to defined models, correct deployment architectures are generated and one of them is selected to be deployed. This generation process is based on a multi-scale modeling approach adopting Bigraphs. In fact, the architecture of a scale is refined by adding the components of the next scale. Then, the obtained architecture is in turn refined and so on, until reaching the last scale. The transition between scales is performed through applying refinement rules. Based on correct by design, the refinement process is executed on a correct scale architecture (respects the defined models) by applying correct rules. In [7] authors propose a cooperative caching scheme for structured data via clusters based on peer connectivity in mobile P2P networks. The proposed scheme reduces data replacement time in the event of changes in topology or cache data replacement using the concept of temp cache. This scheme shows its limits in the multiscale case. P2P networks, are robust systems because it combines the advantages of DHT and unstructured systems, but remains also limited by their drawbacks, such as dependence on resources of the same type(files only) and the exponential number of messages generated during a flood search. Faced with this work, we are motivated to propose the specification of a more generic hierarchical architecture, which allows the storage of all data formats while retaining the advantages in terms of network performance.

3 GRAPP&S

3.1 Model and Definitions

Model and Definitions. For the definition of the GRAPP&S architecture we consider a communication model represented by an undirected and connected graph $G = (V, E)$, where V designates the set of nodes of the system and E designates the set of communication links that exist between the nodes. The model used for our system is studied in [2]. Two nodes u and v are said to be adjacent or neighboring if and only if u, v is a communication link of G. $u_i, v_j \in E$ is a bidirectional channel connected to port i for u and to port j for v. So nodes u and v can send or receive messages to each other in asynchronous mode. A message m in transit is noted $m(id(u), m', id(v))$ or $id(u)$ is the identifier of the node sending the message, id (v) is the identifier of the receiving node, m' indicates the content of the message. Each node u in the system has a unique identifier id and has two primitives: **send(message)** and **receive(message)**. For the sake of clarity, we introduce some definitions.

Definition 1. A node is defined as being a capacity of calculation, of storage, with means and channels of communications.

Definition 2. Raw data is a stream of bytes which can be in different forms: an object database or relational, a file (text and hypertext, XML), a stream (video, audio, VoIP), P2P files, database queries or results from a calculation/service.

Communication and Overlay Networks. The communication model presented in Sect. 3.1 is generic enough not to influence the way whose messages are actually delivered, being limited only to the definition of bidirectional communication properties between two vertices. For this reason, the GRAPP&S architecture can be based on any one overlay communication network which guarantees a reliable two-way communication between two vertices, which allows to explore different communication paths for each directed edge (routed network). This gives greater freedom of implementation and adaptation to the runtime environment, since send/receive operations can be implemented in different ways, depending on the capabilities of node communication. In this case, three main scenarios can be considered:

- Push, where the sender is able to send a message directly to the recipient, - Pull, where the receiver regularly searches for messages on standby (this model is frequently used in the case of networks behind a NAT/firewall), and - Proxy, where the neighbors must go through an intermediary node in order to exchange messages (for example, thanks to a publisher/subscriber middleware). In these three scenarios, it is always possible to establish a direct or partial neighborhood between the processes, which is compatible with the model by directed connected graphs and which therefore meets the needs of the GRAPP&S architecture.

3.2   Elements of the GRAPP&S Architecture

In order to present our architecture, we first introduce some notations. A community (Ci) is an autonomous entity, which groups together nodes which can communicate with each other and which share a defined property: same location, same administration authority (remote servers belonging to the same company, for example) or same application domain (business database, for example). A community contains a single Communicator - (c) process and at least one Resource Manager process- (RM) and a Data Manager - (DM) and these processes are organized hierarchically in a community. The interconnection between different C communities is done through point-to-point neighborhood links between Communicator processes.

Communicator (C). The Communicator node(c) plays an essentially related role the transport of information and the interconnection between different communities, such as when passing messages through firewalls. It is the entry point to the community, and it ensures its security from the outside, through the establishment of Service-Level Agreements (SLAs) with other communities. Likewise, the communicator coordinates the internal security of the community, and can modify its access policies through decisions made within the community [8]. A node c has a unique identifier (ID) from which we construct the identities of the other nodes in the community. This node does not store data and does not index.

Resource Manager (RM). The Resource Manager (RM) processes index and organize data and services in the community. They receive requests from users and ensure their preprocessing. RM nodes participate in finding data in the community. For fault tolerance and performance purposes, the information indexed by the RMs can be redundant and/or partially distributed (DHTs, for example). In order to make the coordination of RMs more efficient, we recommend the election of a designated RM (see Sect. 4.c) ger which allows the node DM to communicate with the RM node to which it is connected.

Data Management (DM). Data Manager (DM) processes interact with data sources, which can be in different media such as databases (object or relational), documents (text/XML/multimedia), streams (video, audio, VoIP), data from sensors or a cloud service. A DM node is a service that has the following components: (i) an interface (proxy) adapted to the different sources data (hard drive, WebDAV server, FTP, database, Dropbox type cloud storage, etc.) and connected to them by a connection protocol specific to the type of data, for example JDBC, ODBC, FTP, etc. (ii) a query manager that allows you to express local or global queries and ((iii) a communication manager which allows the node DM to communicate with the RM node to which it is connected.

3.3 Community Management

GRAPP&S can be deployed in several types of architecture depending on the placement of the nodes. In the placement model (i), the nodes can be grouped together in a single physical machine (see Fig. 1.a). This is a typical example of a machine from a private individual, who wishes to host an architectural community. The placement of nodes in this form can be justified by its simplicity to implement during its implementation phase, using the concepts of inheritance and polymorphism. The nodes are interconnected by sockets, RPC solutions so that they can communicate by message in both directions between two nodes.

GRAPP&S can be deployed in several types of architecture depending on the placement of the nodes. In the placement model (i), the nodes can be grouped together in a single physical machine (see Fig. 1a). This is a typical example of a machine from a private individual, who wishes to host an architectural community. The placement of nodes in this form can be justified by its simplicity to implement during its implementation phase, using the concepts of inheritance and polymorphism. The nodes are interconnected by sockets, RPC solutions so that they can communicate by message in both directions between two nodes.

In (ii) the nodes are organized in a server farm such as a cluster, which is characteristic of HPC networks (Fig. 1.b). Finally, (iii) nodes can be grouped together if they share the same location or administration property (see Fig. 1.c). This is an example of a network formed by the nodes of a company or a research laboratory. Each GRAPP&S node has a unique identifier (ID). IP or MAC addresses are not sufficiently precise identifiers because they do not uniquely identify the different nodes that may reside on the same machine (for example, one RM and several DMs).

Additionally, the use of IP or MAC addresses does not guarantee a unique identity, as private IP addresses can be reused just as much as MAC addresses. Indeed, some unscrupulous manufacturers reuse the MAC addresses which are assigned to them, and this causes many problems in the local networks, just like in the deployment of IPv6 networks. Thus, we propose the solution which consists for each node thus has a single ID_local string, in the form "$urn : communityname : uuid : bit - string$". The expression of the hierarchical addressing is done by the concatenation of the IDs in the form of a prefix, i.e., the ID of the node ci is equivalent to its ID_local, the ID of the ID_{RM_i}

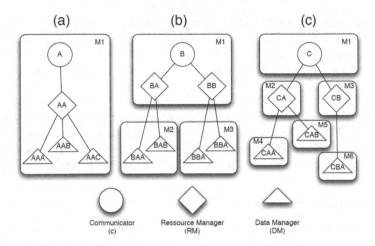

Fig. 1. Organization of nodes (a) in a machine, (b) in a cluster and (c) in a network

node is formed by ID_{c_i}-ID_{RM_i}, and the DM_i node ID has the form ID_{c_i}-ID_{RM_i}-ID_{DM_i}. An advantage of using a clean addressing model to GRAPP&S is that this makes it independent of the addressing model of the overlay network on which GRAPP&S is implemented. Thus, two GRAPP&S communities implemented on different middleware (TomP2P and Phex3, for example) will always be compatible, once the connection has been established between their communicators.

3.4   Node Management

The topology of the network changes frequently due to the mobility of nodes. We are working under the assumption or nodes that arrive in the network are initially a node of type DM. Depending on the conditions of the environment where this node is located, it may be assigned additional roles and "move up" in the hierarchy.

Connecting a Node. When a DM node arrives in the network it has two ways to find an RM node on which it can to log in.

– If the DMi node knows one or more RM nodes, it sends a Hello() broadcast message and collects all the identities of the RM nodes, which it keeps in an array ordered by the identifier. He can thus connect to the RM node which has the highest identifier. If the latter disconnects, then the DMi removes it from the table and connects to the next RM node;
– If on the other hand the DM node does not know any RM node, it must carry out a discovery on the local network (for example, thanks to a multicast) or contact a directory service which can indicate the identifier of a node RMi.

As the way to find the RMi node depends on the implementation, it is not specified in our architecture. Finally, if no attempt to connect to an RM node (and by extension a c node) is successful, the DM node has the option of forming its own community. It thus assumes the three roles c, RM and DM, until other nodes join it. At this time, an election can take place in order to redistribute the roles between the nodes.

Disconnecting a Node. Nodes may experience voluntary disconnections or unintentional (breakdowns). As in the case of disconnections voluntary is trivial, here we focus on unintentional disconnections. Between two hierarchical levels, failures can be detected either by periodic Pull type messages (also known as heartbeat), on demand by Push messages (ping-pong) [9] or even by relying on a specific mechanism. Middleware overlay. For nodes belonging to the same hierarchical level, monitoring can also be done through a mechanism for passing "service" tokens. This not only allows the lightening of the detection mechanism (it is enough to monitor its predecessor and its successor) as allows the rapid dissemination of information to all nodes. For the implementation of a generic failure detection mechanism, we recommend a two-step procedure. First, each node has a list of neighbors $\{N_1, ..., N_n\}$ composed of nodes in direct contact (for example, an RM_i is in contact with its c, its DMs and its neighbors RM_{i-1} and RM_{i+1}. To this list of neighbors is associated a list of timers of wait $\{ta_1, ...ta_n\}$. When no message from node Nk is received until the expiration of time tak, a suspicion of failure is raised and should be verified with a second node which is also in direct contact with the suspect node. Thus, if the suspicion concerns node c, a node RMi questions its direct neighbor RM_{i+1} with a token message initialized to false. If RM_i received a message from node c before the expiration of its tac timeout, RM_{i+1} changes the token value to true and returns the token message to its sender RMi. This means (indirectly) that node c is not disconnected and node RM_i can send a message to node c again. If on the other hand $RM_i + 1$ has not been contacted recently by c, it will forward a token with the value false which, thanks to the passage of the token, will alert all the RM nodes $\{RM_1, .., RM_n\}$ of the failure of c. Similarly, if a node c suspects an RM_k node, it can ask RM_{i+1} for confirmation. Obviously, this generic procedure can be adapted to different situations such as a node which contains one RM and several DMs. In this case, the detection mechanism can be lightened to better respond to the characteristics of the node. Following the confirmation of a failure, the affected nodes must (i) update their information (neighbor list, index tables, etc.) and possibly (ii) proceed to the election of a new RM (respectively c) which will take care of any orphan DM (or RM).

Election Algorithms. Given the dynamic and volatile nature of computer networks, it is important to choose an election algorithm that is as light and responsive as possible. The election of a node may be necessary in two situations:

either to replace a failed node and guarantee continuity of service (for example, during the failure of a node c), but also to simplify coordination between nodes of the same type, with for example the election of an RM which would act as a "supernode" for the indexing of data and services. It should be noted that prior knowledge of higher level nodes is not mandatory, as different techniques are used to obtain the identifiers of other nodes. The simplest method consists in using the GRAPP&S addressing mechanism directly: being independent of the communications middleware, this addressing system makes it easy to go up the GRAPP&S hierarchy and to contact other nodes (thanks to the routing of the r Water overlay). It is therefore sufficient to go up the levels of your own identifier or to contact other nodes whose identifiers have been collected (those from which you have recently received requests, for example). This technique also allows contact other c_j communicators and re-integrate into a network of communities after their ci communicator inadvertently disconnects. As a last resort, GRAPP&S can rely on any topology discovery mechanisms (broadcast/multicast) offered by the own overlay. Since the problem of reconnecting to the rest of the community can be dealt with more or less easily within the GRAPP&S own architecture, it is interesting to look into the election algorithms themselves.

In GRAPP&S, we recommend an election algorithm distributed model inspired by OSPF and IS-IS routing protocols [13]. Indeed, GRAPP&S nodes have a unique identifier that can be used systematically by these election algorithms. The choice between the IS-IS or OSPF algorithms is more related to implementation preferences and node heterogeneity. Indeed, the IS-IS election algorithm is deterministic, where the chosen one is always the node with the largest identifier (called DIS - Designated IS). This mechanism is simple to implement and requires practically no exchange of information because the nodes already have a list with the identifiers of their neighbors, there would only remain the cost associated with taking up the functions of an elected node has a different role from the one it previously occupied. The disadvantage of this technique is that a network with a high rate of volatility can cause repeated elections, either when the leader is disconnected or when a node is connected with a priority identifier.

In cases where volatility may impact the performance of the network, it is possible to use a non-deterministic mechanism such as OSPF. In this type of more conservative algorithm, the choice of a leader (DR - Designated Router) is only necessary if the current leader disappears. Thus, the entry of new nodes into the community has a less significant impact on the functioning of the network.

4   Operations in GRAPP&S

4.1   Storage and Indexing

Data storage in the GRAPP&S network involves Data Manager DM nodes, while Resource Manager RM nodes are used to index data and services. At the end,

each piece of data is uniquely identified by the DM node identifier, to which
is added an extension containing information and the MIME type of the data.
This makes it possible to cross the barrier of the simple "file name", and can
therefore make coexist static data (files), dynamic data (queries on a database,
results of a calculation) and temporary data (voice or video stream, state of a
sensor, etc.).

Adding new data to the network is done as follows: When a DM_i node arrives
in the network, it connects to an RM node and publishes the characteristics of
its data to be indexed. Any changes to the data on a DM are propagated to the
RM to which it is connected, which can then update this information and share
it with other RMs.

This information propagation can take different forms depending on the policies used when implementing the RM network. An implementation that wants
to keep it simple can simply keep a local index on each RM, which will be con-
sulted during a search. On the contrary, an implementation wishing to minimize
the exchanges during a data search will consider the use of a super-node within
the RMs or a DHT mechanism. It is also possible to promote the replication
of indexes and (see data), which requires coordination between RMs in order
to keep copies consistent. In any case, the overload of a node's functionalities
("supernode") is not an obligation in our structure but simply a specificity that
may be present in a given implementation.

4.2   Basic Routing

All of our primitives, whether for resource research or data transfer, are based on
a common and standard routing scheme within our architecture. We can therefore use a hierarchical routing scheme, adapted to GRAPP&S. A community of
GRAPP&S is a tree T, whose vertices correspond to its different components.
If one considers T with a classification of the vertices of T according to a DFS
[12], by construction one thus obtains: for each vertex X of identifier IdX the
address of X is constituted by the binary string representing IdX concatenated
by the binary strings of the parent vertices of X in T. The address of each vertex
X is therefore constituted by a binary string P ATHX and an integer LpathX
representing the length of the chain.

For any vertex X in the tree T, we consider the variable MasqueF the mask
formed by a binary string of LpathX + 8 bits whose first LpathX bits are all at
0 and the last 8 bits are at 1. For example for a vertex X such that LpathX = 16,
its mask will be MasqueX = 00000000. 00000000. 11111111. This construction
allows to obtain the following properties: let Y be a vertex of the tree T, if Y is
a descendant of X in T, then PATHX is a substring of PATHY and by applying
a logical AND between MasqueX = 00000000.00000000.11111111 and PATHY
we have the identifier of Y.

If Y is not a descendant of X in T, then we have to go through the father of X to get to Y. Let Ancestor (X, Y) be the function which for any pair of vertices returns TRUE or FALSE depending on whether X is the parent of Y in the tree T or not. The architecture of GRAPP&S is hierarchical and consists of three levels. This limits the size of the node addresses to a reasonable size, even in a large network.

```
m : message to send to the destination;
local_address : address of the local node;
destination : destination node;
add : Adresse d'un nœud;
Procedure Route(m[, destination])
    If (destination ≠ ∅) then
        If (canReach (destination)) then
            add ← getAdd (destination);
            Send(m, add, local_address);
        else
            add ← getNextHop (local_address, destination);
            Send(m, add, local_address);
        end If
    else
        add ← father;
        Send (m, add, local_address);
    end If
End
```

Algorithm 1: Basic routing method

The basic routing algorithm (see Algorithm 1), allows the transmission of messages between two vertices, knowing the identifier of the source node and that of the destination node, and of course, without having to calculate routing tables. The main functionalities of the GRAPP&S architecture are research and access to data and services. Research in GRAPP&S makes use of hierarchical routing which is based on the identifiers of the nodes (prefixed addressing) which define the paths through which the search requests pass. As we will see below, access to data and services does not depend on a direct connection between nodes. In GRAPP&S, it is always possible to reconcile information by routing in the event that a direct connection between the nodes is impossible.

```
m : message;
role : role of the node (DM, RM, c);
m.type : tmessage type (req — reply —
get — data);
Procedure Receive(m)
  If (role=DM) then
    If (m.type=reply) then
      m'    ←    ("get",    m.data,
      local_address)
      Route(m', m.source);
    end If
    If (m.type = get) then
      m'    ←    ("data",    data,
      local_address)
      Route(m', m.source);
    end If
  end If
  If (role=RM) then
    If (m.type=req) then
      If
      (recherche_locale(m)=FAUX)
      then
        If (origin=child) then
          Route(m, father);
        end If
      end If
    else
      Route(m, m.destination);
    end If
  end If
  If (role=c) then
    If (m.type=req) then
      Diffusion(m);
    else
      Route(m, m.destination);
    end If
  end If
End
```

Algorithm 2: Message processing
according to the roles of the nodes

```
m : search message;
m.source : address of the DM which
sent the message;
m' : reply message;
local_address : local node address;
Function Recherche_locale(m) : boolean
  listeDM ← RM.Verif_of_RM(m.data)
  If (listeDM ≠ ∅) then
    For ( each DMk of RM such as
    DMk ∈ listeDM ) do
      m'←    ("reply",    m.data,
      DMk)
      Route(m', m.source);
    end For
    return TRUE
  else
    return FALSE
  end If
End
```

Algorithm 3: Local search

```
Index_of_RM[0...n] : index du nœud RM;
X ←{} : list of DM;
TypeMime : MIME type of information ;
TypeMimeRec : Search criteria;
Function Verif_of_RM(TypeMimeRec) : X
←{}
  For (i = 0 to n) do
    If (Index
    _of_RM[i].TypeMime=TypeMimeRec)
    then
      X ← DMi;
    end If
  end For
  return X;
End
```

Algorithm 4: Verifying the index RM

```
m : message to be broadcast;
origin : address of the RM which
transmitted the message;
local_address : : address of the local
node;

Procedure Diffusion(m, origin)
  For ( each child RMk of ci such as
  RMk ≠ origin) do
    Route(m, RMk)
  end For
End
```

Algorithm 5: Message broadcast
to RMs

4.3   Data Search

Among the main functionalities of GRAPP&S is search and access to data. This section describes our algorithm for finding data in a community of GRAPP&S. The different nodes which intervene in this research have some basic functions, some exposed previously, on which we will rely to define our algorithms: Send allows a source node to send a message to a destination node. However, this primitive requires a direct connection between nodes, which is why it is associated with the Route primitive; Route allows a source node to transmit a message to any node, either by sending it directly through Send, or by relaying the information to a node with the routing mechanism prefixed.

In order to get an address in the prefixed routing mechanism, we use a get-NextHop() method which performs the comparison of the addresses and applies the subnet masks in order to get the address of the next node towards the destination. Receive which is used to process the messages received, according to their types and according to the role played by the node (see Algorithm 3). $Local_research$ that allows a DM to respond to a resource search. When searching, messages can be of two types:

- type search message: ("Req", content, source)
- type response message: ("Reply", content, source)

When a client searches for data on GRAPP&S, it comes into contact with a DM_i proxy, which sends a Y request containing the characteristics of the data and the type req. So the Y query takes the form < "req", content, source>. As the message does not have a specific recipient, research in a GRAPP&S community is done in stages, respecting the hierarchical organization of the network. The search procedure is as follows:

1. $DM_i \in C_i$ sends the request Y to its node $RM_i \in C_i$ using Route () (see Algorithm 1).
2. RM_i receives the request Y according to its type (see Algorithm 3) and checks using the $Local\_Search$ (m) method (see Algorithm 1) whether in its index there is a neighbor DM which contains the data sought
3. if yes, then the RM_i node returns to the DM_i node a list of nodes DM which contain the information sought, in the form of a Y'' message with the type reply. The form of the Y'' request is ("Reply", content, source). The illustration is given in Fig. 6 where the node $DM_i \in C_i$ with identifier AAA sends the Y request of type req to its node $RM_i \in C_i$ using the Route() function. The AA identifier RM_i node replies by sending a reply type message containing the identifier of the DM_y node which contains the data sought.
4. if not, the node RM_i forwards the request to its node $c_i \in C_i$ for a broadcast to the other nodes $RM_i \in C_i$ such that the $RMk \neq RMi$ (Algorithm 5). Each of these RM_k nodes execute the $Local_search(m)$ method (see Algorithm 1) to check if in its local index there is a neighbor DM which contains the data sought,

5. When an $RM_i \in C_i$ node finds a match, a reply stamped reply from this message with the form < "reply", content, source> will be returned to the sender DM_i node by taking the reverse path to go using Route () (see Algorithm 1).
6. If none of the $RM_k \in C_i$ nodes finds a match, and if the C_i community is connected to other communities, then the ci forwards the request according to the access policies defined and implemented, or negotiated, in place in each community.

At this search step, we do a flood search, flooding on the community tree. This search by flood is illustrated in Fig. 2 where the communicator node Ci of identification A broadcasts in the community Ci to all $RM_k \in C_i$ the message Y of type Req. The identifier RM_k node $BB \in C_j$ finds a match and responds to the AAA identifier DM_i node by taking the reverse path with a reply type message containing the identifier of the DM_y node that owns the data. However, it can be noted that the flooding is limited to the structure of the GRAPP&S communities, and not to the structuring of the underlying peer-to-peer network. Here are now the different algorithms of the functions presented previously. First, the specific management of messages according to the different types of nodes. Then the Local Search for a resource algorithm. Now the Resource Manager index test and verification algorithm. Finally, the algorithm ensuring the dissemination of messages to the Resource Manager.

4.4   Data Transfer in GRAPP&S

If the search is successful, the client (DM_x) obtains the identifier of the DM_y node that holds the data. During the data transfer stage, messages can take two different forms

- content request message $(get, content, source)$
- content transfer message $(data, content, source)$

When a node receives a message of type $(data, content, source)$, if it is not the target of this message, then it retransmits the message. If the latter is intended for him then he decides either to display the content of the message or to store it thanks to a Save() primitive on a support such as: on disk, on a cloud service such as Dropbox or Seafile, on a messaging service such as Gmail, etc. In this case, he has two possibilities to contact DM_y and recover the data: (i) by direct connection; (ii) by a routed connection if direct connection is not possible. In both cases, we will call the Route() primitive. If direct connection is possible, sending will be done by $Send()$, otherwise it will be done by the hierarchical routing mechanism. $Remark\ 1.$ As the resource was found by a search and its location was returned to the site that initiated the request, there is therefore a bidirectional path using the links of the hierarchical architecture of GRAPP&S, which can be used between the requester and the resource.

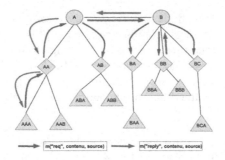

Fig. 2. Research in the GRAPP&S network

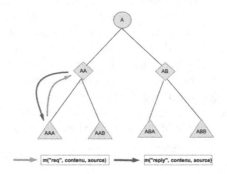

Fig. 3. Local search in GRAPP&S

Direct Connection Possible. If a direct connection is possible between the requesting node and the owner of the resource, the DM_x node directly sends the request $(get, content, source)$ to the DMy node by a simple Send() method (see Algorithm 2). The latter DM_y responds with the transfer message $((data, content, source)$ by following the reverse path forward (see Algorithm 2). This is the example of nodes that are in the same subnet. An illustration is given in Fig. 3 where the AAA identifier DM_x node sends directly to the node AAB identifier DM_y, a request to access the data. The latter responds by following the reverse path. Then the DM_x node can view the data or store it on a medium such as a disk, on a cloud service, on Gmail etc., thanks to the save() function. *Remark 2.* Two ABC and DEF identifier nodes may well be part of the same network, and be part, as their identifiers show, of two GRAPP&S communities. In this case, a direct connection will then be possible. For direct connection not possible.

Direct Connection is Not Possible. If direct connection is not possible, then the message is sent by hierarchical routing in GRAPP&S. The hierarchical routing mechanism is applied, and therefore the resource will follow the path taken by the search request and its associated response. In this case, the operations carried out are as follows

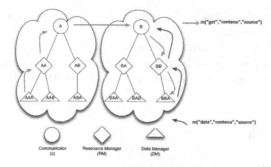

Fig. 4. Data repatriation via a routed connection

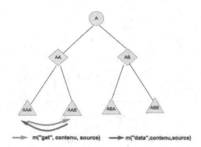

Fig. 5. Repatriation in the case of a local network in GRAPP&S

- the DM_x client sends a get type request to its RM, which will retransmit the get message using its Route() primitive to its parent, the communicator node C_i.
- the C_i node, according to the prefixed routing (see Algorithm 2), sends the get request to the RM node which indexed the data.
- The latter, by prefixed routing (see Algorithm 2), forwards the get request to the DM_y node responsible for the data.
- the DM_y node, which owns the data, can transmit it with the Route () primitive (see Algorithm 2), thanks to a data type message by following the reverse path on the way. Then the DM_x node can visualize the data or store it on a support arbitrary, thanks to the save() function.

Figure 4 Illustrates this routed connection, where the AAA Identifier DM_x node sends the get request message using its Route() primitive that goes up the tree to download the data managed by the BBA Identifier DM_y node. The latter the DM_y node in the same way, when it receives a get type message, the DM_y node which holds the data can transmit it with the Route() primitive, thanks to a data type message (Fig. 5).

This hierarchical search mechanism prevents the flooding of network links. The hierarchy allows to define the paths by which the requests transit and as the logical connectivity of our architecture is by definition of (n − 1), it suffices to apply a PIF type algorithm to aggregate the requests and reduce the number of messages from a search.

As access to GRAPP&S data does not depend on a direct connection between nodes, it is always possible by routing to retrieve the data by the path used during the search, in case a direct connection between the nodes does not is not possible.

5   Future Work

5.1   Development of a Prototype on Pastry

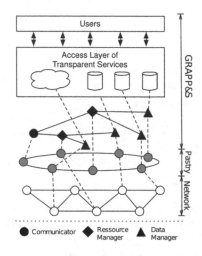

Fig. 6. Illustration of the distribution of nodes GRAPP&S on a Pastry overlay

A prototype of the GRAPP&S architecture is under development. This prototype uses the TomP2P overlay network to interconnect the various nodes of the GRAPP&S architecture, as shown in Fig. 6. A prototype of the GRAPP&S architecture is under development. This prototype uses the TomP2P overlay network to interconnect the various nodes of the GRAPP&S architecture, as shown in Fig. 6. The use of TomP2P is purely practical because the Pastry [11] overlay takes care of any low level interconnection (opening of sockets, detection of failures, etc.), thus allowing developers to concentrate on the coordination of GRAPP&S nodes (election, indexing of data and services, processing of requests, research, etc.). The use of a design pattern like Facade allows a modular development where the overlay can be easily replaced. Indeed, the performance constraints related to node management and data transfer may be too high for TomP2P, if the community is expected to have a high number of nodes or if the network has a high level of volatility. In addition to demonstrating the application of the GRAPP&S architecture and the implementation of scalability tests, this prototype will also allow the analysis of the impact of the location of the nodes in relation to the management of the hierarchy. Indeed, the operation of GRAPP&S is based on a hierarchical structure for the routing, the search and the interconnection of remote nodes. If the placement of nodes has an impact on the performance of the architecture, it should also be taken into account when selecting RMs and c.

5.2   Resource Redistribution with GAIA

While GRAPP&S initially acts as a mediator for the location of resources stored in the different DMs, it also has the potential for development thanks to the redistribution of resources through the GAIA scheduler. Like the body of the same name for the work of Isaac Asimov, GAIA aims to integrate and possibly reorganize resources between the different MDs, according to criteria such as frequency of use, proximity e vis-à-vis customers, the need for replication, the cost of storage and, of course, the feasibility of such a distribution. Indeed, the GAIA project concerns the development of a scheduler that will be able to manage the storage of certain types of data in order to increase the efficiency of access to the most used elements, while reducing the storage costs of less critical data. Thanks to GAIA, GRAPP&S will be able to better manage the integration of "slow" storage services such as Amazon Glacier, which present a reduced cost but also more restrictive access policies than traditional storage mechanisms.

5.3   Future Application of Our Architecture

Our architecture will be used to connect the health institutes (public for example), interested in the sharing of resources through a pooling of the health data of each institute in the form of "community". Indeed the revolution in information technologies, and the spread of the Internet of Things (IoT) and smart city systems, have fostered widespread use of smart systems. As a complex, no stop service, healthcare requires efficient and reliable follow-up on daily operations, service and resources.

Faced with the challenges of cloud-only sharing solutions where data storage is remote and access speed slow, Cloud and edge computing are essential for smart and efficient healthcare systems in smart cities. Our framework will become a decentralized edge computing solution that will allow rapid access to resources due to the proximity of data and consumers.

6   Conclusions

In this article we present our work around GRAPP&S (GRid Applications and Services), a multi-scale architecture for data aggregation and services. We present the main features of GRAPP&S, such as the specification of its components, connection and disconnection mechanisms, indexing operations, research and access to data, as well as the principles recommended for the coordination of knots. Based on the principles of multi-scale systems GRAPP&S was designed as a hierarchical network based around the concept of "communities", which allows the integration of information sources with heterogeneous access protocols and various security rules. In addition, the use of Data Managers, specialized proxies for the adaptation and processing of different types of data, it is possible to transparently integrate both file type data as well as databases, flows (audio, video), web services and distributed computation. GRAPP&S is a work in progress, including a prototype in development course will be used for passing tests scale and as a platform for advanced services such as security grids [7] and storage optimization with the GAIA scheduler, introduced in this article.

References

1. Rottenberg, S., Leriche, S., Taconet, C., Lecocq, C., Desprats, T.: MuSCa: a multiscale characterization framework for complex distributed systems (2014)
2. Chalopin, J., Godard, E., Métivier, Y., Ossamy, R.: Mobile agent algorithms versus message passing algorithms. In: Shvartsman, M.M.A.A. (ed.) OPODIS 2006. LNCS, vol. 4305, pp. 187–201. Springer, Heidelberg (2006). https://doi.org/10.1007/11945529_14
3. Satyanarayanan, M.: Mobile computing: the next decade. SIGMOBILE Mob. Comput. Commun. Rev. **15**, 2–10 (2011)
4. Gassara, A., Rodriguez, I.B.: Describing correct deployment architectures based on a bigraphical multi-scale modeling approach. Comput. Electr. Eng. **63**, 277–288 (2017)
5. Jiang, C., Gao, L., Duan, L., Huang, J.: Scalable mobile crowdsensing via peer-to-peer data sharing. IEEE Trans. Mob. Comput. **17**(4), 898–912 (2018)
6. Lee, D., Park, N., Kim, G., et al.: De-identification of metering data for smart grid personal security in intelligent CCTV-based P2P cloud computing environment. Peer-to-Peer Netw. Appl. **11**, 1299–1308 (2018)
7. Bok, K., Kim, J., Yoo, J.: Cooperative caching for efficient data search in mobile P2P networks. Wirel. Pers. Commun. **97**, 4087–4109 (2017)
8. Huang, X., Qin, Z., Liu, H.: A survey on power grid cyber security: from component-wise vulnerability assessment to system-wide impact analysis. IEEE Access **6**, 69023–69035 (2018)
9. Jeanneau, D., Rodrigues, L., Arantes, L., et al.: An autonomic hierarchical reliable broadcast protocol for asynchronous distributed systems with failure detection. J. Braz. Comput. Soc. **23**, 15 (2017)
10. Roy, D.G., De, D., Mukherjee, A., et al.: Application-aware cloudlet selection for computation offloading in multi-cloudlet environment. J. Supercomput. **73**, 1672–1690 (2017)
11. Rowstron, A., Druschel, P.: Pastry: scalable, distributed object location and routing for large-scale peer-to-peer systems. In: IFIP/ACM International Conference on Distributed Systems Platforms (Middleware), pp. 329–350, November 2001
12. Fraigniaud, P., Gavoille, C.: Routing in trees. In: Orejas, F., Spirakis, P.G., van Leeuwen, J. (eds.) ICALP 2001. LNCS, vol. 2076, pp. 757–772. Springer, Heidelberg (2001). https://doi.org/10.1007/3-540-48224-5_62
13. Bhatia, M., Manral, V., Ohara, Y.: IS-IS and OSPF Difference Discussion. IETF Internet Draft, January 2006
14. Satyanarayanan, M., Bahl, P., Caceres, R., Davies, N.: The case for VM-based cloudlets in mobile computing. IEEE Pervasive Comput. **8**, 14–23 (2009)

A Leading Edge Detection Algorithm for UWB Based Indoor Positioning Systems

Sreenivasulu Pala$^{(\boxtimes)}$ and Dhanesh G. Kurup

RF and Wireless Systems Laboratory, Department of Electronics and Communication Engineering, Amrita School of Engineering, Amrita Vishwa Vidyapeetham, Bengaluru, India
{p_sreenivasulu,dg_kurup}@blr.amrita.edu

Abstract. Ultra-wide band (UWB) based indoor positioning systems are preferred for accurate positioning applications due to the high time resolution offered by UWB signals. This paper proposes an algorithm for determining the leading edge of UWB signal for indoor positioning applications and assess its performance in different signal-to-noise ratio (SNR) conditions. We found that the proposed algorithm outperforms the previously published methods, especially under low SNR situations. The performance of the algorithm can be further increased with an additional number of anchor nodes.

Keywords: UWB · Indoor positioning system · Leading-edge detection

1 Introduction

Indoor positioning systems (IPS) have different applications such as asset tracking, indoor way finding, space management and location-based services. IPS offers benefits such as efficiency improvement, convenience and cost saving. IPS uses technologies like Bluetooth, Wi-Fi, RFID and UWB for position estimation [1–6]. UWB technology offers high positioning accuracy compared to other technologies due the high bandwidth of UWB signals [7–9]. IPS employs various ranging techniques such as received signal strength indicator (RSSI), time of arrival (TOA), time difference of arrival (TDOA) and angle of arrival (AOA) for target position estimation. RSSI based techniques determine the distance between target node and anchor node using RSSI [10] whereas AOA based techniques estimate the angle between target node and anchor node using antenna arrays [11,12]. The RSSI based techniques are sensitive to the channel parameters and hence are less accurate. The AOA based methods do not perform well in multi-path environments [13]. TOA based techniques use the signal propagation delay for range estimation and need proper synchronization between the target and anchor nodes. TDOA approach uses TOA difference at each pair

N. Kumar et al. (Eds.): UBICNET 2021, LNICST 383, pp. 45–55, 2021.
https://doi.org/10.1007/978-3-030-79276-3_4

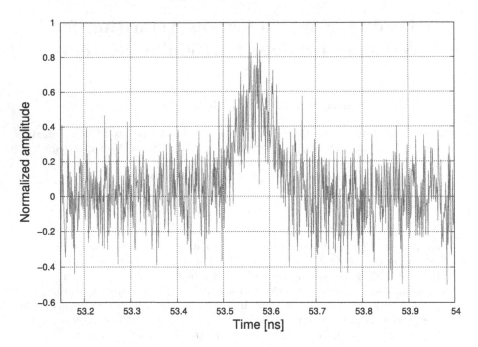

Fig. 1. Received signal at SNR = 4 dB

of anchor node for target position estimation [14,15] and requires synchronization only among the anchor nodes. UWB based IPS utilizes TOA or TDOA based ranging method for position estimation. The anchor node of an UWB IPS implements a TOA estimation algorithm for range estimation. Different UWB receiver structures and TOA estimation algorithms are proposed in the literature [16,17]. A Single channel receiver that implements standard leading edge detection (SLED) algorithm [19,20] provides an improved accuracy compared to other methods [18]. The SLED algorithm is used to detect the leading edge in the received UWB signal. However, the SLED algorithm fails to detect the leading edge in low SNR conditions.

In this paper, we propose an Improved Leading Edge Detection (ILED) algorithm and evaluate its performance using simulations. The results show that the proposed algorithm outperforms the original SLED algorithm [19], especially under high noise conditions. The rest of the paper is organized as follows. Section 2 describes the proposed ILED algorithm, trilateration method and simulation setup of UWB localization system. Section 3 describes the simulation results concerning accuracy of the proposed algorithm and influence of the number of anchor nodes on the positioning accuracy. Finally, Sect. 4 concludes the paper.

2   System Model

In an UWB based IPS, the target node transmits an UWB pulse $g(t)$ to different anchor nodes given by,

$$g(t) = \frac{A}{\sqrt{2\pi}\sigma} e^{-\frac{t^2}{2\sigma^2}} \tag{1}$$

where A is the amplitude of the Gaussian pulse and σ is the shaping factor. The signal received at an anchor node $h(t)$ can be represented as

$$h(t) = g(t) + a(t) \tag{2}$$

where $a(t)$ represents Additive White Gaussian Noise (AWGN).

2.1   Proposed ILED Algorithm

The received signal $h(t)$ at an SNR of 4 dB is shown in Fig. 1. As shown in the figure, the received signal $h(t)$ is fully noisy and it is difficult to detect the leading edge of the transmitted Gaussian pulse $g(t)$ in it. We have improved the existing SLED algorithm to develop the proposed ILED algorithm for detecting the leading edge in such noisy signals. The original SLED algorithm [19, 20] and improvements to it are described below.

SLED Algorithm

At first, a Z point moving average filter is applied on the received signal $h(t)$ to obtain $z(t)$ where a Z point moving average filter returns the average of previous Z samples. In the next step, two maximum window filters of size F_1 and F_2 are applied on $z(t)$ to obtain signals $f_1(t)$ and $f_2(t)$ respectively, where $F_2 > F_1$. Here, the F point maximum window filter returns the maximum-valued sample in the previous F samples.

The leading edge of $h(t)$ is detected when the below condition (3) is fulfilled where η is the leading edge factor.

$$\frac{f_2(t)}{f_1(t)} \leq \eta \tag{3}$$

In order to avoid mis-detection of the leading edge in pure noise, the leading edge is detected only when UWB signal is present in the received signal. Hence, the condition (3) is checked only when a signal $v(t)$ crosses a threshold w given by,

$$v(t) \geq w \tag{4}$$

where,

$$w = G_{noise} + t_{swt} \tag{5}$$

$$v(t) = f_2(t) \tag{6}$$

Here, G_{noise} is the maximum value of filtered received signal $z(t)$ before the occurrence of the UWB pulse and t_{swt} is a pre-defined threshold [20].

Improvements to SLED Algorithm

The original SLED algorithm provides poor performance for SNR < 6 dB, due to false detection of the leading edge in noise due to steps (5)–(6) [20]. Since the standard deviation of noise $\sqrt{\tilde{G}}$ is a consistent measure of noise, we use it in our algorithm as noise threshold in (4), instead of the maximum value of filtered noise as in the case of original algorithm (5). The probability of false detection of leading edge increases due to the usage of $f_2(t)$ in (6) in the original algorithm. This is because $f_2(t)$ crosses the noise threshold before the onset of UWB pulse due to $F_2 > F_1$. In order to reduce the chance of mis-detection of leading edge, a faster signal shall be used in (4). Modified leading edge detection (MLED) algorithm [21] implements faster maximum window filter signal $f_1(t)$ for noise threshold comparison, as mentioned in (7).

$$v(t) = f_1(t) \tag{7}$$

The accuracy of the algorithm can be further improved by using the fast moving average signal $z(t)$ for noise threshold comparison instead of faster maximum window filter signal $f_1(t)$. This is due to the reason that $z(t)$ leads in time when compared to $f_1(t)$ and hence by using $z(t)$, the chance of mis-detection of the leading edge in noise can be reduced. Hence, the following changes are implemented when compared to the SLED and MLED algorithm to improve the performance in high noise conditions.

$$w = \gamma\sqrt{\tilde{G}} \tag{8}$$

$$v(t) = z(t) \tag{9}$$

where $\tilde{G}$ indicates the noise power calculated from $h(t)$ in noise, before the occurrence of UWB pulse and γ is the factor of noise standard deviation. Lower values of γ increases the chance of detecting leading edge in noise prior to UWB pulse. On the other hand, higher values of γ miss the leading edge detection under high noise conditions. Hence, by choosing an optimum value for γ by means of simulations or experiments, an optimum performance can be achieved using the proposed algorithm.

2.2   Trilateration

Once the leading edge is detected in the signal received at an anchor node, the distance between the target node and the respective anchor node can be determined. Further, the target node position can be determined using the distance between target node to each anchor node based on trilateration as explained below.

Consider an IPS with N number of anchor nodes located at (x_i, y_i) where i indicates the anchor node index and a target node located at (x_t, y_t). Assume the distance from target node to each anchor node is r_i. Using the Euclidean distance between target and anchor nodes, the following non-linear system of equations can be obtained.

$$\left.\begin{array}{ll} (x_0 - x_t)^2 + (y_0 - y_t)^2 & = \hat{r}_0{}^2 \\ (x_1 - x_t)^2 + (y_1 - y_t)^2 & = \hat{r}_1{}^2 \\ \quad\vdots \\ (x_{N-1} - x_t)^2 + (y_{N-1} - y_t)^2 = \hat{r}_{N-1}{}^2 \end{array}\right\} \tag{10}$$

The following over-determined linear system of equations can be obtained by solving the above non-linear equations.

$$\mathbf{Bx} = \mathbf{c} \tag{11}$$

where,

$$\mathbf{B} = \begin{bmatrix} (x_0 - x_1) & (y_0 - y_1) \\ \vdots & \vdots \\ (x_0 - x_{N-1}) & (y_0 - y_{N-1}) \\ \vdots & \vdots \\ (x_{N-2} - x_{N-1}) & (y_{N-2} - y_{N-1}) \end{bmatrix} \quad \mathbf{c} = \begin{bmatrix} C_{01} \\ \vdots \\ C_{0(N-1)} \\ \vdots \\ C_{(N-2)(N-1)} \end{bmatrix}$$

$$C_{ij} = \frac{1}{2}.(x_i^2 - x_j^2 + y_i^2 - y_j^2 + \hat{r}_j{}^2 - \hat{r}_i{}^2)$$

$$\mathbf{x} = \begin{bmatrix} x_t & y_t \end{bmatrix}^{\mathrm{T}}$$

The target node position, $\hat{\mathbf{x}} = [\hat{x}_t \ \hat{y}_t]^{\mathrm{T}}$ can be estimated using the least square solution of (11) given by,

$$\hat{\mathbf{x}} = (\mathbf{B}^{\mathrm{T}}\mathbf{B})^{-1}\mathbf{B}^{\mathrm{T}}\mathbf{c} \tag{12}$$

2.3 System Simulation

An UWB IPS simulator is developed that supports a minimum of 3 and a maximum of 6 anchor nodes. The anchor nodes are located on the vertices of a regular polygon with a radius of 13 m. The location of the target node is chosen inside the circumference of a circle that encloses the polygon vertices. By choosing such a setup, the target node gets optimum coverage from all the anchor nodes. An UWB signal $g(t)$ with a pulse width of 300 ps is used for the simulations. The received signal $h(t)$ is modeled according to (2) where the noise power is

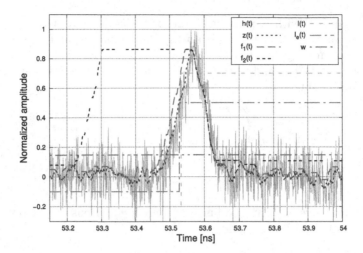

Fig. 2. Illustration of signals involved in the proposed leading edge detection algorithm for SNR = 10 dB and tuning parameters, $\eta = 1.2$ and $\gamma = 1.2$

determined based on the simulated SNR. The proposed algorithm with $Z = 16$, $F_1 = 16$ and $F_2 = 256$ is applied on $h(t)$ to detect the leading edge. Finally, the position of the target is estimated using (12). Figure 2 shows all the signals involved in ILED algorithm which includes, the received signal $h(t)$, 16 point moving averaged signal $z(t)$, 16 point and 256 point maximum window filtered signals $f_1(t)$ and $f_2(t)$ respectively, the estimated leading edge binary signal $l(t)$, expected leading edge binary signal $l_e(t)$ and the noise threshold w. As discussed in Sect. 2.1, we can clearly see from the Fig. 2 that the 16 point moving average filter output signal $z(t)$ leads in time when compared to the 16 point maximum window filter output $f_1(t)$. By using $z(t)$ for noise threshold comparison, we can avoid leading edge detection in noise, prior to the onset of UWB pulse.

3   Results

3.1   Ranging Accuracy

Monte-Carlo simulations are performed with the target position fixed at the center of polygon, to estimate the distance between the anchor node and the target node using the proposed algorithm. Ranging accuracy $E^{(1D)}$ is evaluated as mentioned below in (13).

$$E^{(1D)} = \sqrt{\frac{1}{N_T.N_{BS}} \sum_{i=1}^{N_T} \sum_{j=1}^{N_{BS}} \left[r_{err}^{ij} \right]^2} \qquad (13)$$

where,

$$r_{err}^{ij} = \sqrt{(x_t - x_a^j)^2 + (y_t - y_a^j)^2}$$
$$- \sqrt{(\hat{x}_t^i - x_a^j)^2 + (\hat{y}_t^i - y_a^j)^2},$$

N_T and N_{BS} represents the number of trials and number of anchor nodes respectively, (x_a^j, y_a^j) represents the position of jth anchor node and (x_t, y_t), $(\hat{x}_t^i, \hat{y}_t^i)$ represents the actual and estimated target positions during the ith trial, respectively.

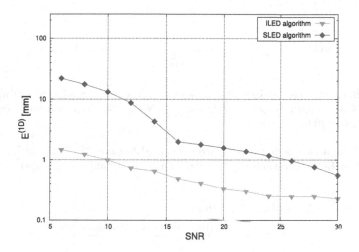

Fig. 3. Performance comparison of proposed ILED algorithm with SLED algorithm under normal SNR conditions for $\eta = 1.2$, $\gamma = 1.2$, $N_T = 100$, $N_{BS} = 3$ and target node position fixed at (0,0)

Figure 3 shows the $E^{(1D)}$ of the proposed algorithm and SLED algorithm for different normal SNR conditions. As we can see from Fig. 3, the proposed algorithm clearly provides less error when compared to the SLED algorithm. The proposed algorithm reports sub-centimeter accuracy for SNR in between 6 dB and 10 dB and sub-millimeter accuracy for SNR with in 10 dB and 30 dB, while the SLED algorithm reports an accuracy of higher than 1 cm for 6 dB < SNR < 10 dB and sub-centimeter accuracy for 10 dB < SNR < 25 dB.

Figure 4 shows the comparison of ranging accuracy between the proposed algorithm and SLED algorithm, under low SNR conditions. As we can see from Fig. 4, the SLED algorithm is unable to detect the leading edge for SNR < 1 dB, and hence reports a very high error, while the proposed algorithm reports sub-centimeter accuracy for this SNR range. For 0 dB < SNR < 6 dB, the SLED algorithm reports an error of higher than 2 cm, while the proposed ILED algorithm reports sub-centimeter accuracy. Table 1 reports the positioning error for the complete range of SNR.

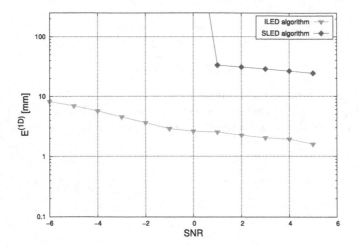

Fig. 4. Performance comparison of proposed ILED algorithm with SLED algorithm under low SNR conditions for $\eta = 1.2$, $\gamma = 1.2$, $N_T = 100$, $N_{BS} = 3$ and target node position fixed at (0,0)

Table 1. Comparison of positioning error using SLED and ILED algorithms for different SNR in LOS conditions. 'x' indicates an unacceptably high positioning error.

SNR (dB)	$E^{(1D)}$ (mm)	
	SLED Algorithm	ILED Algorithm
−6	x	8.2
−2	x	3.6
2	30.8	2.2
6	22.0	1.4
10	13.1	0.9
14	4.3	0.6
18	1.7	0.4
22	1.3	0.3
26	0.9	0.2
30	0.5	0.2

From these results, we can conclude that the proposed algorithm provides improved accuracy when compared to the SLED algorithm. Especially under low SNR conditions, we can clearly see that the proposed algorithm outperforms the original algorithm.

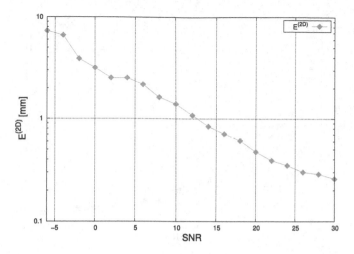

Fig. 5. $E^{(2D)}$ for $\eta = 1.2$, $\gamma = 1.2$, $N_T = 100$, $N_{BS} = 3$ and target node position is chosen randomly

3.2 Positioning Accuracy

Monte-Carlo Simulations are performed with target position chosen randomly inside the circumscribed circle of polygon. The target position is estimated using the proposed algorithm and the trilateration described in Sect. 2.2. Positioning accuracy $E^{(2D)}$ is evaluated as mentioned below in (14).

$$E^{(2D)} = \sqrt{\frac{1}{N_T} \sum_{i=1}^{N_T} \left[(x_{err}^i)^2 + (y_{err}^i)^2 \right]} \qquad (14)$$

where,

$$x_{err}^i = (x_t^i - \hat{x}_t^i), \quad y_{err}^i = (y_t^i - \hat{y}_t^i)$$

Figure 5 reports the positioning accuracy for different SNR values using 3 anchor nodes. As we can see from Fig. 5, the proposed system reports sub-centimeter accuracy for SNR < 12 dB and sub-millimeter accuracy for 13 dB < SNR < 30 dB. From these results, we can conclude that the proposed ILED algorithm in combination with the least square based trilateration provides high positioning accuracy.

3.3 Influence of Number of Anchor Nodes on the Positioning Accuracy

In order to study the influence of the number of anchor nodes on the positioning accuracy, $E^{(2D)}$ is determined for 3 to 6 number of anchor nodes. Simulations are performed with target positions chosen randomly inside the circumference of a circle that encloses the polygon vertices. Figure 6 depicts the positioning

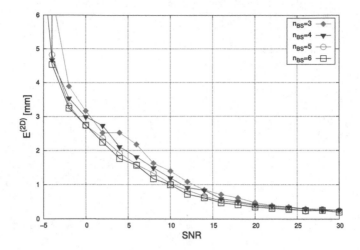

Fig. 6. Influence of number of anchor nodes on 2D positioning accuracy for $\eta = 1.2$, $\gamma = 1.2$ and the target position is randomly chosen

accuracy for different SNRs and number of anchor nodes. From the figure we can note that the positional accuracy improves with the increase in the number of anchor nodes for each SNR. This is due to an increased coverage of the target by more anchor nodes. Hence, we can conclude that the positioning accuracy of the proposed algorithm can be improved with an additional number of anchor nodes.

4   Conclusion

We propose an algorithm for determining the leading edge of a received UWB signal under Additive White Gaussian Noise (AWGN) conditions for indoor positioning applications. The proposed algorithm estimates accurate target position even under low signal-to-noise ratio (SNR) conditions. The accuracy of the proposed algorithm can be increased by using an additional number of anchor nodes. The proposed method finds its use in precise indoor positioning applications.

References

1. Yang, C., Shao, H.R.: WiFi-based indoor positioning. IEEE Commun. Mag. **53**(3), 150–157 (2015)
2. Jianyong, Z., Haiyong, L., Zili, C., Zhaohui, L.: RSSI based bluetooth low energy indoor positioning. In: 2014 International Conference on Indoor Positioning and Indoor Navigation, October, pp. 526–533 (2014)
3. Pala, S., Palliyani, S., Himdi, M., Lafond, O., Kurup, D.G.: Localization of unknown electromagnetic source using 3D-antenna arrays. Int. J. Microw. Wirel. Technol. **12**, 86–94 (2019)

4. Chandrasekaran, V., Narayan, K., Vasani, R.K., Balasubramanian, V.: InPLaCE RFID: indoor path loss translation for object localization in cluttered environments. In: 2015 IEEE 10th International Conference Intelligent Sensors, Sensor Networks and Information Processing, ISSNIP 2015, April, pp. 7–9 (2015)
5. Milioris, D., Tzagkarakis, G., Papakonstantinou, A., Papadopouli, M., Tsakalides, P.: Low-dimensional signal-strength fingerprint-based positioning in wireless LANs. Ad Hoc Netw. **12**, 100–114 (2014)
6. Malla, H., Purushothaman, P., Rajan, S.V., Balasubramanian, V.: Object level mapping of an indoor environment using RFID, pp. 203–212 (2014)
7. Zebra Technologies Corp.: Dart UWB Technology Datasheet (2012). https://www.zebra.com/
8. DecaWave: DW1000 IEEE802.15.4-2011 UWB Transceiver DataSheet (2011). https://www.decawave.com/
9. Time Domain: Data Sheet PulsON® 440, pp. 1–64 (2015). https://www.timedomain.com/
10. Luo, X., O'Brien, W.J., Julien, C.L.: Comparative evaluation of received signal-strength index (RSSI) based indoor localization techniques for construction jobsites. Adv. Eng. Inform. **25**(2), 355–363 (2011)
11. Tay, B., Liu, W., Zhang, D.H.: Indoor angle of arrival positioning using biased estimation. In: IEEE International Conference on Industrial Informatics, pp. 458–463 (2009)
12. Zhang, Y., Brown, A.K., Member, S., Malik, W.Q., Edwards, D.J.: High resolution 3-D angle of arrival determination for indoor UWB multipath propagation. IEEE Trans. Wirel. Commun. **7**(8), 3047–3055 (2008)
13. Gezici, S., et al.: Localization via ultra-wideband radios: a look at positioning aspects of future sensor networks. IEEE Signal Process. Mag. **22**(4), 70–84 (2005)
14. Zhang, J., Dong, F., Feng, G., Shen, C.: Analysis of the NLOS channel environment of TDOA multiple algorithms. In: 2015 IEEE SENSORS, Busan, pp. 1–4 (2015)
15. Monica, S., Ferrari, G.: UWB-based localization in large indoor scenarios: optimized placement of anchor nodes. IEEE Trans. Aerosp. Electron. Syst. **51**(2), 987–999 (2015)
16. Hernandez, A., Badorrey, R., Chóliz, J., Alastruey, I., Valdovinos, A.: Accurate indoor wireless location with IR UWB systems a performance evaluation of joint receiver structures and TOA based mechanism. IEEE Trans. Consum. Electron. **54**(2), 381–389 (2008)
17. Mahfouz, M.R., Zhang, C., Merkl, B.C., Kuhn, M.J., Fathy, A.E.: Investigation of high-accuracy indoor 3-D positioning using UWB technology. IEEE Trans. Microw. Theory Tech. **56**(6), 1316–1330 (2008)
18. Zhang, C., Kuhn, M.J., Merkl, B.C., Fathy, A.E., Mahfouz, M.R.: Real-time noncoherent UWB positioning radar with millimeter range accuracy: theory and experiment. IEEE Trans. Microw. Theory Tech. **58**(1), 9–20 (2010)
19. Merkl, B.C.: The future of the operating room: surgical preplanning and navigation using high accuracy ultra-wideband positioning and advanced bone measurement (2008)
20. Kuhn, M.J., Turnmire, J., Mahfouz, M.R., Fathy, A.E.: Adaptive leading-edge detection in UWB indoor localization. In: 2010 IEEE Radio and Wireless Symposium, RWW 2010 - Pap. Dig., pp. 268–271 (2010)
21. Pala, S., Jayan, S., Kurup, D.G.: An accurate UWB based localization system using modified leading edge detection algorithm. Ad Hoc Netw. **97**, 102017 (2020). https://doi.org/10.1016/J.ADHOC.2019.102017

Designing Multiband Millimeter Wave Antenna for 5G and Beyond

Tirumalasetty Sri Sai Apoorva$^{(\boxtimes)}$ and Navin Kumar🆔

Department of Electronics and Communication Engineering, Amrita School of Engineering,
Amrita Vishwa Vidyapeetham, Bengaluru, India

Abstract. In this work, we have presented microstrip low profile patch antenna at Millmeter wave (mmWave) for 5G and future communication. There are two designs presented in this work. The first, we designed a dual band in 5G new radio (NR) specified frequency bands of 27 GHz and 37 GHz. We achieved the design with maximum gain of 8 dB. In the second design, we modified the structure to obtain multiband; again, in the International Telecommunication Union (ITU) specified possible bands for mobile communication. The design was optimized for 5 different bands that is, 28 GHz, 35.5 GHz, 41 GHz, 51 GHz and 60 GHz. We obtained reasonably good bandwidth in each band. The basic patch dimension is 12.5×10.1 mm^2. In order to increase the gain of the basic element, the array of antenna is designed. Linear array with element up to 10 is designed and simulated. It is observed that an increase in the gain of around 3 dB gain is achieved. On optimization, it is found that with 6 element array the performance is relatively better on both basic bands.

Keywords: MmWave antenna · Multiband mmWave antenna · 5G mobile communication · Array antenna · 27 GHz and 37 GHz

1 Introduction

5G networks are being globally deployed at rapid pace. This 5G wireless communications and network is design to provide fiber like experience to users [1, 2]. Many new techniques like massive multiple input multiple output (MIMO), advanced coding, and mobile millimeter wave (mmWave) have been developed and being investigated. At the same time, network capacity is expected to be increased approximately 100x. To expand the network capacity, 5G new radio (NR) air interface enables diverse spectrum in both, the sub-6 GHz and mmWave frequency bands [3]. This additional spectrum required by 5G brings new challenges for product design and global deployment, especially in the mmWave frequency range [4]. To achieve high data rates assured by 5G, the antenna system becomes a crucial component in the overall system design.

Additionally, higher frequency bands in mmWave have been proposed in 5G and backhaul network [5]. Therefore, the requirements of multiband antenna especially in these higher bands and with significant bandwidth are increasing. Furthermore, a dual

© ICST Institute for Computer Sciences, Social Informatics and Telecommunications Engineering 2021
Published by Springer Nature Switzerland AG 2021. All Rights Reserved
N. Kumar et al. (Eds.): UBICNET 2021, LNICST 383, pp. 56–65, 2021.
https://doi.org/10.1007/978-3-030-79276-3_5

band multi-band operation with polarization diversity finds useful applications in current and the next generation of wireless systems [6]. Not only that, a dual-band antenna with different I/O ports is also preferable in many practical applications. However, most of the integrated designs work at a single band and polarization [7]. And, due to the asymmetrical structure of the radiation elements, most of the designs are not appropriate for dual-polarization applications. Also, the gain from one single antenna is found to be limited from 3–6 dB. High gain and directivity requirements can be obtained by array of antenna elements where individual radiating beams of each antenna are pooled to provide a high directivity. Authors in [8], designed and discussed a dual frequency quarterwave shorted microstrip patch antenna for satellite MIMO application to achieve relative higher gain.

Several efforts are made in antenna geometries as well as array configurations at microwave and mmWave frequencies including at 28- and 38 GHz [9, 10] to provide improved gain and bandwidth. In many reported antennas design, the desired performance is linked with design complexities usually because of multilayer structures. Such design leads to high fabrication cost. Moreover, mmWave antenna design is mostly focused on planar configurations. However for 360° area coverage, integrated array on four faces of a cube with two antennas on each face [11] are used. Yet in another approach [12], multi-faceted phased array is investigated. In this, conformal microstrip antenna arrays are presented for continuous conformal surfaces at 35- and 32 GHz. In [13], different multiband antenna configurations for 28/38 GHz have been demonstrated for 5G applications. A dual band 27/37 GHz with approximately 3 GHz bandwidth and over 8 dB gain is also presented in [14]. However, most of the works are dual band or 3-band with limited gain and bandwidth. In [15], authors tried to design a slot antenna for wider bandwidth. Additionally, in lower band of frequencies such as sub-6 GHz, there are multiple designs available with multiband [16–18].

In this work, we present a design of 5-band mmWave antenna. The bands are: 27 GHz, 35.5 GHz, 41 GHz, 51 GHz and 60 GHz. The bandwidth in each case is more than 1.3 GHz and the gain varies from over 6 dB to 8 dB. This is interesting results. In order to increase the gain, we also designed linear array of antennas but with basic dual band antenna. For the array, the feed is given only for the first element, remaining patches act as the passive elements. The passive elements get the fringing effect + the radiation of previous element as feed and then they radiate. Number of elements was tried in the array to ensure optimum gain on both the frequency 27/37 GHz band. It is observed that the array with 6-elements offer approximately 1.5 dB gain in the lower band while close to 3 dB gain in higher band.

Rest of the contents in the paper is arranged as follows. Section 2 presents the design of antenna while results are presented in Sect. 3. In Sect. 4, the conclusion is given.

2 Design of Antenna

2.1 Theoretical Calculation of the Basic Patch

Initially, a micro strip patch antenna was designed at 27 GHz of frequency for which width and length is calculated. The width of the patch is given as:

$$w = \frac{c}{2fr}\left(\sqrt{\frac{2}{\varepsilon_r + 1}}\right) \tag{1}$$

where fr is the centre frequency and ε_r is the permittivity. However, normally effective permittivity is used to avoid fringing effect present during the design. It is given as:

$$\varepsilon_{eff} = \frac{\varepsilon_r + 1}{2} + \frac{\varepsilon_r - 1}{2}\left(\frac{1}{\sqrt{1 + \frac{2h}{w}}}\right) \tag{2}$$

where, h is the height of the substrate. The length is calculated as the effective length by including the resonant frequency and the effective relative permittivity. The total length is calculated as the $L_{eff} - 2\Delta L$.

$$L_{eff} = \frac{c}{2fr\sqrt{\varepsilon_{eff}}} \tag{3}$$

$$\Delta L = h \times 0.412 \times \frac{\left(\varepsilon_{eff} + 0.3\right)\left(\frac{w}{h} + 0.264\right)}{\left(\varepsilon_{eff} - 0.258\right)\left(\frac{w}{h} + 0.8\right)} \tag{4}$$

$$L = L_{eff} - 2\Delta L \tag{5}$$

The basic structure of the antenna is developed by using the above calculation.

2.2 Dual Band Design

Initially, a dual band millimeter wave antenna is designed to operate in the millimeter wave centered at frequency 27.5 GHz and 35.7 GHz [10]. The thickness of the designed antenna is 0.87mm and the substrate material used is Rogers Duroid 5880 (RT Duroid 5880). The relative permittivity is 2.2. The strip line feeding technique is used to feed the antenna structure. The design is shown in Fig. 1 and the corresponding dimensions are shown in Table 1.

2.3 Multiband Design

Several changes in structure were tried to get bands without degrading the obtained parameters from the basic element (Fig. 1). In the process, we duplicated the basic antenna and connected through a rectangular patch of length 13.05 mm that is, 13.05 mm apart as shown in Fig. 2. The patch width is 0.6mm and position of the patch from the feed is 2.2 mm. This structure offered 3rd band at 41 GHz. We continued the design with many other structural changes but then found more bands when another rectangular patch at the center is connected. However, with only centered connected patch (as shown in Fig. 3), we obtained 4-bands with additional band at 60 GHz. Such approach encouraged us to find more operating bands. Our objectives were to get multiple bands in the operating frequency up to 62 GHz. We then kept two of the rectangular patches (centre and edge) as shown in Fig. 4. With optimized distances and dimensions, we obtained 5-bands of interest. Important dimensions are labeled in the Table 1 itself.

Table 1. Dimensions of the single element Antenna

Dimension	Value
L1	6.7
L2	8.1
L3	2.2
L4	6.7
L5	2.2
L6	0.2
L7	0.7
L8	1.9
W1	0.6
W2	2.9
W3	0.9
W4	1.4
W5	1.5
a	0.2

2.4 Antenna Array Design

To obtain higher gain, we created array of antenna. For this, we started with basic elements and its array. That is, dual band (Fig. 1 design) with higher gain. The spacing between the antenna elements is $\lambda/2$ i.e., 5.45 mm from the edge of the T-structure and the same distance is maintained for all the elements. The antenna array has a series fed structure. The first antenna element is feed by using a coaxial feed technique. The elements next to the first element will receive the radiation from the first, of course, may be with some propagation delay. Multiple elements are linearly connected to offer more gain. We obtained maximum gain at both frequency bands. On optimization, it is found that with 6 and 10 elements array, the gain is approximately the same.

Slightly decrease in the gain of lower band with 10-element array is observed. Therefore, we concluded that 6-element array performs optimal in this type of feeding. The array with 10-element is shown in Fig. 5.

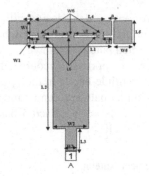

Fig. 1. Dual band antenna

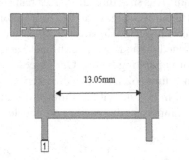

Fig. 2. Antenna array with edge connection that gave the 3rd band

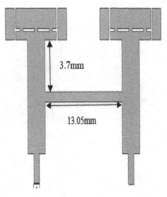

Fig. 3. Antenna element design that produced the 4th band

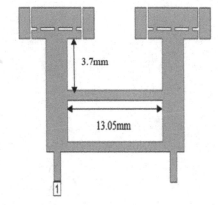

Fig. 4. Multiband (5) antenna elements that produced 5 bands

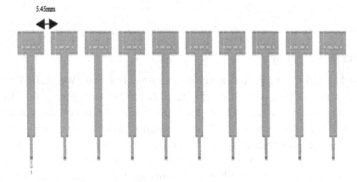

Fig. 5. Antenna array with the basic elements

3 Results and Discussion

As mentioned before, we used design in applied wave research software (AWR). Figure 6 illustrates S11 parameter of 3-bands from the design of Fig. 2 (Antenna array with edge connection that gave the 3rd band). Though, it was not an objective, however, it is also found to be of practical use. We see the 3-bands at 27 GHz, 37 GHz and 41 GHz. The S11-parameters for the three bands are −14.35 dB, −23.52 dB and −49.61 dB which are reasonable. The gain plot for different frequencies is shown in Fig. 7. Figure 7(a) shows the gain at 27.8 GHz which is 8.66 bdB. The gain at 35.8 GHz is 9.707 dB and that of at 41.1 GHz is 6.185 dB are shown in Fig. 7(b) and Fig. 7(c) respectively. They are also comparable, in fact, better than most of the designs considering single element multiband.

The S11-parameter of the antenna design of Fig. 3 (Antenna element design that produced the 4th band) is shown in Fig. 8. It is to be noted that this design has resulted into 4-bands with the last being at 60 GHz.

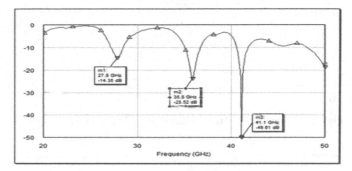

Fig. 6. S11 parameters for edge design

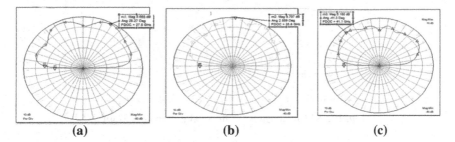

(a)   (b)   (c)

Fig. 7. (a) Gain at 27.8 GHz (b) Gain at 35.8 GHz (c) Gain at 41.1 GHz

The frequency bands are 28 GHz, 36 GHz, 41 GHz, 59 GHz and their respective reflection coefficients are −11.4 dB, −16.6 dB, −2029 dB and −22.11 dB. The gain is also in the range over 6 dB to 9 dB as shown in Fig. 9(a)–(c).

The final multiband design (Fig. 4) has resulted in a 5-band where the center frequencies and the reflection coefficients are at 28 GHz, S11 is −18.53 dB; at 36 GHz,

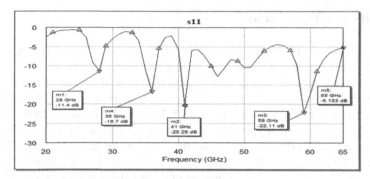

Fig. 8. S11 of 4-bands from design

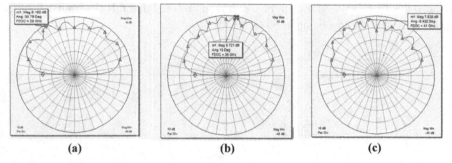

(a)   **(b)**   **(c)**

Fig. 9. (a) Gain at 28 GHz (b) Gain at 36 GHz (c) Gain at 41 GHz

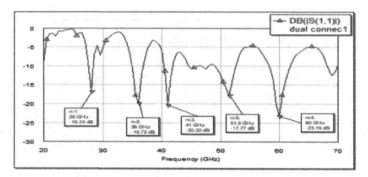

Fig. 10. S11 parameters for the final multiband design

S11 is −19.72 dB; at 41 GHz, the reflection coefficient is −20.38 dB; at 51.5 GHz, S11 is −21.94 dB and, finally at 60 GHz, S11 is −21.1 dB which is shown in Fig. 10. Similarly, the gains remain intact in the range 6.89 dB to 9.6 dB in all the bands. The corresponding gains at 51.5 GHz is 10.98 dB and at 60 GHz the gain is 9.788 dB which is shown in Fig. 11(a) and (b).

We also simulated the performance of array design as shown in Fig. 5. With 6-element array, the performance on both bands is found to optimum. The center frequencies of the

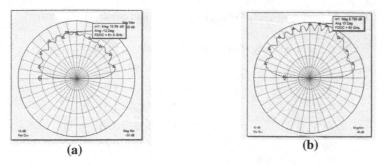

(a) **(b)**

Fig. 11. (a) Gain at 51.5 GHz (b) Gain at 60 GHz

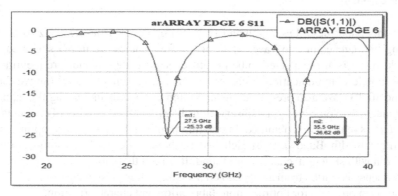

Fig. 12. S11 parameters for Array design

array with passive elements are 27.5 GHz and 35.5 GHz and, the reflection coefficients as −25.33 dB and −26.62 dB is shown in Fig. 12. The corresponding gain at 27.5 GHz is 7.739 dB and at 35.5 GHz it is 9.924 dB. The same result is summarized in Table 2.

Table 2. Summary of the final design of the multi band antenna design

Number of antenna elements	Center frequency (GHz)	S11 (dB)	Gain (dB)
6	27.5	−25.33	7.816
	35.5	−26.62	10.6

Finally, overall results are listed in Table 3 for final design of 5-band mmWave antenna. We observed that a single antenna can be designed to operate in wide band in mmWave from 28–60 GHz with good bandwidth and gain over 8 dB.

Table 3. Summary of the final design of the multi band antenna design

Center frequency (GHz)	S11(dB)	Gain	Bandwidth
28	−16.74	8.192	1.23
36	−19.72	9.721	1.91
41	−20.32	7.838	1.51
51.5	−17.77	10.58	3.81
60	−23.19	9.788	3.62

4 Conclusion

Work in this paper describes multiband (5-band) operating from 27 GHz to 60 GHz operating frequencies in mmWave. Most of the frequencies are in 5G specified band and backhaul. It is also expected to be in future usage as these bands are identified by ITU for mobile communication. Achieved gains are reasonably good (above 7.8 dB to around 10 dB). We also presented our first attempt towards increasing the gain with array of antennas. Some important observations like maximum number of elements with passive feed is realized and obtained. In creating multiband, we have also optimized the gain and bandwidth. Bandwidth in each case is above 1.3 GHz. Interesting results and observation with structural changes are found. It is seen that more bands might result if basic frequency is started from around 40 GHz instead of what we have taken as 27 GHz. Our work is in progress for optimization, fabrication and characterization.

References

1. Wang, C.X., et al.: Cellular architecture and key technologies for 5G wireless communication networks. IEEE Commun. Mag. **52**(2), 122–130 (2014)
2. Gupta, A., Jha, R.K.: A survey of 5G network: architecture and emerging technologies. IEEE Access **3**, 1206–1232 (2015)
3. ETSI TS 138 104 V15.2.0 (2018-07) 3GPP, Technical Specifications: 5G: NR; Base Station (BS) radio transmission and reception, July 2018
4. Sheeba Kumari, M., Rao, S.A., Kumar, N.: Characterization of mmWave link for outdoor communications in 5G networks, 10–13 Aug 2015
5. Jaber, M., Imran, M.A., Tafazolli, R., Tukmanov, A.: 5G backhaul challenges and emerging research directions: a survey. IEEE Access **4**, 1743–1766 (2016). https://doi.org/10.1109/ACCESS.2016.2556011
6. Jung, Y.-B., Eom, S.-Y.: A compact multiband and dual-polarized mobile base-station antenna using optimal array structure. Int. J. Antenna Propag. **2015**, 1–12 (2015)
7. Karamzadeh, S., Kartal, M., Saygin, H., Rafii, V.: Polarisation diversity cavity back reconfigurable array antenna for C-band application. IET Microwaves Antennas Propag. **10**(9), 955–960 (2016)
8. Sindhu, B., Jayakumar, M.: Design and analysis of dual frequency quarterwave shorted microstrip patch antenna for satellite MIMO. In: International Conference on Wireless Communications Signal Processing and Networking (WiSPNET), 21–23 March 2019

9. Hong, W., Baek, K., Ko, S.: Millimeter-wave 5G antennas for smartphones: overview and experimental demonstration. IEEE Trans. Antennas Propag. 65(12), 6250–6261 (2017)
10. Hasan, M.N., Bashir, S., Chu, S.: Dual band omnidirectional millimeter wave antenna for 5G communications. J. Electromagn. Waves Appl. 33(12), 1581–2159 (2019)
11. Cheng, Y.J., Xu, H., Ma, D., Wu, J., Wang, L., Fan, Y.: Millimeter-wave shaped-beam substrate integrated conformal array antenna. IEEE Trans. Antennas Propag. 61(9), 4558–4566 (2013)
12. Khalifa, I., Vaughan, R.: Optimal configuration of multi-faceted phased arrays for wide angle coverage. In: Vehicular Technology Conference (VTC-2007), Spring (2007)
13. Wang, Q., Ning, M., Wang, L., Safavi-Naeini, S., Liu, J.: 5G MIMO conformal microstrip antenna design. Wirel. Commun. Mobile Comput. 2017, 1–11 (2017). Article ID 7616825
14. Sri Sai Apoorva, T., Kumar, N.: Design of mmWave dual band antenna for 5G wireless. In: IEEE International Conference on Advanced Networks and Telecommunications Systems, India, 16–19 December 2019
15. Arvind, V., Sowmiya, R., Vignesh, P., Kavin, S., Jayakumar, M.: Yagi slot antenna based on substrate integrated waveguide cavity for wider bandwidth applications. In: 2016 International Conference on Advanced Communication Control and Computing Technologies (ICACCCT), 25–27 May 2016
16. Snehalatha, T.K.A.C., Kumar, N.: Design of multiband planar antenna. In: IEEE International Conference on Antenna Innovations & Modern Technologies for Ground, Aircraft and Satellite Applications (iAIM), 26–27 November 2017
17. Hong, Y., Tak, J., Baek, J., Myeong, B., Choi, J.: Design of a multiband antenna for LTE/GSM/UMTS band operation. Int. J. Antennas Propag. 2014, 1–9 (2014). Article ID 548160
18. Snehalatha, T.K.A.C., Kumar, N.: Design of multiband planar antenna for mobile devices. In: IEEE International conference on Microelectronic Devices, Circuits and Systems (ICMDCS), 10–12 August 2017

Analog Beamforming mm-Wave Two User Non-Orthogonal Multiple Access

S. Sumathi[1]([⊠])(iD), T. K. Ramesh[1], and Zhiguo Ding[2]

[1] Department of Electronics and Communication Engineering,
Amrita School of Engineering, Amrita Vishwa Vidyapeetham, Bengaluru, India
{s_sumathi,tk_ramesh}@blr.amrita.edu
[2] Department of Electrical and Electronic Engineering,
The University of Manchester, Manchester, UK
zhiguo.ding@manchester.ac.uk

Abstract. In this paper, the authors aim to design analog beam forming weight vector for two user down link non-orthogonal multiple access (NOMA) scenario. Millimeter wave (mm-wave) channel with one line-of-sight (LOS) path and multiple non line-of-sight (NLOS) paths is considered. The components of beamforming vector vary only in phase whereas their magnitudes are same. This restriction guarantees that power amplifiers need not be designed with different power amplification factors, thereby minimizing the complexity of power amplifier design. In this work, two users located at different angle of departure (AoD) with different gains are considered. The weight vector is designed aiming to minimize the total power requirement. Simulation results show that the proposed approach requires less power with minimum complexity compared with the time division multiple access (TDMA) for the same spectral efficiency requirements.

Keywords: NOMA · mm-wave channel · Analog beamforming ·
Power minimization · Spectral efficiency · Semi-definite relaxation

1 Introduction

Fifth Generation broadband wireless networks and device to device communication require to establish epidemic hike in the traffic volumes and emerging challenges in user data rate. This can be achieved by the integration of enormous spectrum with a powerful radio access technology [1–3]. Huge bandwidth (30 GHz to 300 GHz) of mm-wave is a crucial factor of using this in 5G mobile network to cater multi-gigabit communication services [4,5].

NOMA is one of the essential practices to cope with the enormous data rate necessities and support massive connectivity in 5G networks and beyond [6]. In the existing 4G orthogonal multiple access (OMA) strategies, orthogonal resources such as time, frequency and code are allotted to diverse users to get rid

© ICST Institute for Computer Sciences, Social Informatics and Telecommunications Engineering 2021
Published by Springer Nature Switzerland AG 2021. All Rights Reserved
N. Kumar et al. (Eds.): UBICNET 2021, LNICST 383, pp. 66–76, 2021.
https://doi.org/10.1007/978-3-030-79276-3_6

of inter-user interference. Conversely, same resource block with different power levels can be used for multiple users in power domain NOMA. Far user is allotted more power, near user is allotted less power and Superposition Coding (SC) is employed at the transmitter. Consequently, at the receiver of near user, successive interference cancellation (SIC) is applied to decode the far user's data. Then it is subtracted from the near user's received signal to decode its own signal [7,8]. Hence NOMA can be combined with mm-wave band to meet the requirements of future generation networks.

However, severe propagation losses occur in mm-wave carrier frequencies. These losses can be mitigated by the use of array antenna with beam forming technique. It is also feasible to deploy large number of antenna elements due to very small wavelength of mm-wave band [9]. Beam forming is a method in which phases of antenna elements are dynamically varied to produce a narrow beam [10]. This beamforming increases array gain which in turn enhances the signal-to-noise ratio (SNR) and thereby diminishes the propagation path loss.

Beam forming techniques of NOMA in Rayleigh channel is discussed in [11], where power minimization and fairness based beamforming is achieved for perfect channel state information (CSI) condition. In addition to that, robust scheme is discussed under imperfect CSI, targeting to minimize the transmit power. Authors of [12] considered only two users with digital beamforming. They solved power minimization problem with QoS constraints using semi definite relaxation (SDR) as discussed in [13,14].

In [15,16], random steering of single beam forming method is used. This is applicable only when the NOMA users are closer. When the angle between two users is larger, a single wide beam used will reduce the beam gain and in turn decreases the data rate. Hence NOMA with multi beams each targeting towards NOMA users is an efficient method to achieve high data rate. Authors of [17] also use multi beam NOMA with analog beamforming. In this beamforming, single RF chain is connected to number of antenna elements which reduces the hardware complexity and cost. Here, the users which have different AODs and different gains are served by NOMA with analog beamforming weight vector where the absolute value of all the elements of the weight vector is a constant. They discussed how to allocate power among the two NOMA users and design weight vector to boost the sum rate. They solved this non-convex optimization problem in two steps. However, complexity of this algorithm is high for more number of users. Authors of [18] discusses the reduction in achievable data rate for two user two beam mm-wave NOMA under imperfect CSI case. To the best of our knowledge, analog beamforming to achieve the minimization of total power to meet the target spectral efficiency of both the users has not been studied for mm-wave NOMA.

Contribution of our work is as follows:

We consider two user down link mm-wave multiple input single output (MISO) NOMA system. Analog beamforming vector for power minimization problem is solved in our low complex proposed method which is described as follows: We first transform the problem into semi-definite programming (SDP) problem.

Constant Modulus constraint of analog beamforming vector and rank constraint of positive semi-definite (PSD) matrix are relaxed. Problem is solved to compute the unit norm beamforming vector. Then, normalization is done on each component of weight vector. It is again multiplied with a suitable scalar such that all the constraints are satisfied. Obtained analog beamforming vector is the optimum one since the rank of the PSD matrix is one. It is shown that the total power required in our proposed NOMA is less than the conventional TDMA approach.

Rest of the paper is structured as follows. Section 2 illustrates the model of mm-wave channel. Section 3 describes two user analog beamforming down link mm-wave MISO-NOMA system. Proposed algorithm for computing analog beamforming vector to solve power minimization problem is also presented in Sect. 3. Simulation results are discussed in Sect. 4. Conclusion is offered in Sect. 5. Following notations are used. Lower case and upper case bold face letters are adopted for vectors and matrices respectively. Hermitian, transpose, expectation, absolute operation and Euclidean norm operation are denoted by $(.)^H$, $(.)^T$ $E(.)$, $|.|$ and $\|.\|_2$ respectively.

2   mm-wave Channel Model

The channel characteristics of mm-wave LOS and NLOS directional outdoor links in various reference cases are investigated in [19]. The mm-wave channel $\mathbf{h}_j$ between the base station (BS) and j -th user in the downlink direction as expressed in [20] is given in Eq. (1).

$$\mathbf{h}_j = \sqrt{\frac{v_j}{v_j + 1}}\mathbf{a}_{j,0} + \sqrt{\frac{1}{v_j + 1}}\sqrt{\frac{1}{L}}\sum_{p=1}^{L}\alpha_{j,p}\,\mathbf{a}_{j,p} \qquad (1)$$

First term in (1) represents the LOS path and the second term denotes L number of NLOS scattered paths, where v_j is the Rician factor of j-th user, which is the power ratio between the LOS component and scattered components.

$\alpha_{j,p} \sim \mathbb{C}\,\mathcal{N}\left(0, \sigma_j^2\right)$ where σ_j^2 is average channel power gain of user j. $\mathbf{a}_{j,p}$ is given by

$$\mathbf{a}_{j,p} = \left[1,\ e^{i\pi 1\cos\theta_{j,p}}, ...,\ e^{i\pi(N-1)\cos\theta_{j,p}}\right]^T \qquad (2)$$

Spacing between N numbers of antenna elements in Uniform Linear Array (ULA) antenna is $\lambda/2$. $\theta_{j,p}$ is the AoD of p-th path of user j. $\theta_{j,p}$ is uniformly distributed over $[0, \pi]$. We assume user 1 is closer to BS compared to user 2. Therefore, $\|\mathbf{h}_1\|_2^2 \geq \|\mathbf{h}_2\|_2^2$.

3   System Model and Problem Formulation

3.1   System Model

Two user mm-wave down link MISO NOMA system with analog beamforming weight vector is considered in Fig. 1.

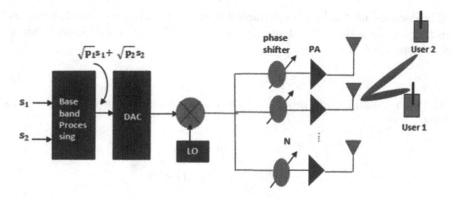

Fig. 1. System model for Two User Analog beamforming down link mm-wave MISO NOMA

BS having N element ULA antenna serves 2 users, each with a single antenna. Each antenna element branch has a phase shifter and a power amplifier (PA). BS produces two concentrated beams pointing towards two NOMA users concurrently. It is achieved by the proper design of phase shifts. Assuming s_j is the symbol involved for user j and p_j is the power required for user j with $\mathbb{E}\{|s_j|^2\} = 1$, and $\mathbf{w} \in \mathbb{C}^{N \times 1}$ be the complex beam forming unit normalized weight vector. Signal received by j-th user is

$$y_j = \mathbf{h}_j^H \mathbf{w} \left(\sqrt{p_1} s_1 + \sqrt{p_2} s_2 \right) + n_j \ , \ j = 1,2 \tag{3}$$

where $p_1 + p_2 = P$, P is the total power transmitted by BS. n_j is the additive white Gaussian noise (AWGN) at j-th downlink user with unit variance $n_j \sim \mathbb{C}\mathcal{N}(0,1)$. $\mathbf{h}_j$ is the mm-wave channel from BS to user j.

3.2 Problem Formulation

We assume perfect CSI is available at the BS. Signal to Interference plus Noise Ratio (SINR) required to decode user 2 data at user 2 is given in (4), where user 1 signal is the interference.

$$\text{SINR}_{2,2} = \frac{p_2 |\mathbf{h}_2^H \mathbf{w}|^2}{p_1 |\mathbf{h}_2^H \mathbf{w}|^2 + 1} \tag{4}$$

At the near user (user 1), SIC is performed. Therefore SINR required to decode user 2 data at user 1 is mentioned in (5).

$$\text{SINR}_{2,1} = \frac{p_2 |\mathbf{h}_1^H \mathbf{w}|^2}{p_1 |\mathbf{h}_1^H \mathbf{w}|^2 + 1} \tag{5}$$

The decoded user 2 data is subtracted from the received SC coded signal and then user 1 data is decoded. Hence the requisite SINR to decode user 1 data at user 1 is specified in (6).

$$\mathrm{SINR}_{1,1} = p_1 \left| \mathbf{h}_1^H \mathbf{w} \right|^2 \tag{6}$$

Total Power minimization problem with analog beamforming is formulated as follows.

$$\min_{p_1, p_2, \mathbf{w}} (p_1 + p_2) \tag{7a}$$

$$s.t. \log_2 (1 + \mathrm{SINR}_{1,1}) \geq r_1 \tag{7b}$$

$$\log_2 (1 + \mathrm{SINR}_{2,1}) \geq r_2 \tag{7c}$$

$$\log_2 (1 + \mathrm{SINR}_{2,2}) \geq r_2 \tag{7d}$$

$$\|\mathbf{w}\|_2^2 = 1 \tag{7e}$$

$$\left| [\mathbf{w}]_n \right| = \text{constant}, \ n = 1, 2, ...N \tag{7f}$$

where r_1 and r_2 are the target Spectral Efficiency (SE) of user 1 and user 2 respectively. Non-convex constraint (7f) is removed and then problem (7) is transformed into SDP as mentioned in problem (8).

$$\min_{p_1, p_2, \mathbf{Q}} (p_1 + p_2) \tag{8a}$$

$$s.t. \ p_1 tr \left(\mathbf{h}_1^H \mathbf{Q} \mathbf{h}_1 \right) \geq \beta_1 \tag{8b}$$

$$(p_2 - \beta_2 p_1) \, tr \left(\mathbf{h}_1^H \mathbf{Q} \mathbf{h}_1 \right) \geq \beta_2 \tag{8c}$$

$$(p_2 - \beta_2 p_1) \, tr \left(\mathbf{h}_2^H \mathbf{Q} \mathbf{h}_2 \right) \geq \beta_2 \tag{8d}$$

$$tr(\mathbf{Q}) = 1 \tag{8e}$$

$$rank(\mathbf{Q}) = 1 \tag{8f}$$

$$\mathbf{Q} \succeq 0 \tag{8g}$$

$$where \ \mathbf{Q} = \mathbf{w} \mathbf{w}^H \ \& \ \beta_j = 2^{r_j} - 1, \ j = 1, 2 \tag{8h}$$

Constraint (8f) is relaxed and the resultant SDR problem is with 2 scalar variables p_1 & p_2 and a PSD matrix $\mathbf{Q}$ with 4 constraints. Rank of $\mathbf{Q}$ in the problem is computed by referring [14].

$$rank^2(\mathbf{Q}) \leq (4 - 2) \tag{9}$$

Hence it is confirmed that the rank of $\mathbf{D}$ is 1 and the solution obtained is always optimal. The optimum beamforming vector $\mathbf{w}^{opt}$ can be obtained from a rank-one $\mathbf{Q}^{opt}$ solution, as

$$\mathbf{w}^{opt} = \sqrt{u^{opt}} \ \mathbf{v}^{opt} \tag{10}$$

where u^{opt} and v^{opt} are the eigenvalue and the eigenvector of $\mathbf{Q}^{opt}$. Solution obtained satisfies $\|\mathbf{w}^{opt}\|_2^2 = 1$. But the elements of the obtained $\mathbf{w}^{opt}$ do not have constant modulus. Hence normalization is done on $\mathbf{w}^{opt}$ as mentioned in

(11) and then multiplied with α such that all the constraints (7b), (7c) & (7d) are satisfied. Let us call the final solution as $\mathbf{w}^{\text{final}}$.

$$[\mathbf{w}_{NM}]_i = \frac{[\mathbf{w}^{opt}]_i}{\sqrt{N}} \frac{}{|[\mathbf{w}^{opt}]_i|}, \quad i = 1, 2, ...N \tag{11}$$

$$\mathbf{w}^{\text{final}} = \alpha \, \mathbf{w}_{NM} \tag{12}$$

where α is a positive scaling coefficient.

The achieved SE of user 1 and user 2 are computed from (13) and (14).

$$R_1 = \log_2\left(1 + p_1\left|\mathbf{h}_1^H\mathbf{w}^{\text{final}}\right|^2\right) \tag{13}$$

$$R_2 = \min\left\{\log_2\left(1 + \frac{p_2\left|\mathbf{h}_1^H\mathbf{w}^{\text{final}}\right|^2}{p_1\left|\mathbf{h}_1^H\mathbf{w}^{\text{final}}\right|^2 + 1}\right), \quad \log_2\left(1 + \frac{p_2\left|\mathbf{h}_2^H\mathbf{w}^{\text{final}}\right|^2}{p_1\left|\mathbf{h}_2^H\mathbf{w}^{\text{final}}\right|^2 + 1}\right)\right\} \tag{14}$$

The required total transmit power of NOMA users is given in (15).

$$P = \left\|\mathbf{w}^{\text{final}}\right\|_2^2 (p_1 + p_2) \tag{15}$$

Performance of our proposed algorithm is compared with the conventional 2 user TDMA approach. Achieved SE of user k in TDMA is specified in (16).

$$R_{k-\text{TDMA}} = \frac{1}{2} \log_2\left(1 + \left|\mathbf{h}_k^H\mathbf{w}_{k-\text{TDMA}}\right|^2\right), \quad k = 1, 2 \tag{16}$$

where the term $\frac{1}{2}$ represents the time slot allotted for each user. $\mathbf{w}_{k-\text{TDMA}}$ is designed to satisfy the maximal ratio combining (MRC) as well as the target SE. It is assumed that the AWGN noise has unit variance.

4   Simulation Results

Simulation has been performed in cvx-matlab [21] for 1000 runs. We assume channel variances of user 1 and user 2 are $\sigma_1^2 = 2$, $\sigma_2^2 = 1$; $N = 32$, $L = 3$. $v_1 = v_2 = 32$; $\theta_{j,p}$ is uniformly distributed over $[0, \pi]$ for $j = 1,2$ and $p = 1,2,...L$. For attaining user fairness, it is assumed $r_1 = r_2$.

Figure 2 and Fig. 3 show achieved SE versus target SE of near user (user 1) and far user (user 2) respectively. Since the common weight vector for both the users is multiplied with a positive constant, achieved SE of near user is more than the target SE. However far user's achieved SE is same as the target SE. In conventional TDMA, achieved SE is same as target SE in near user as well as in far user. This is due to the fact that the weight vector is designed for a particular user in that dedicated time slot. Figure 4 depicts achieved sum SE vs. target sum SE.

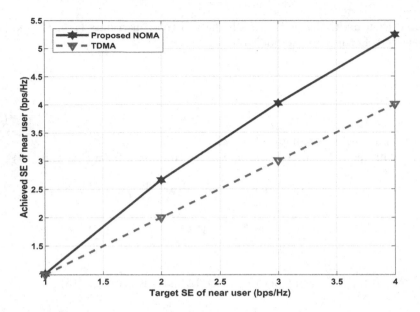

Fig. 2. Achieved SE of near user against target SE

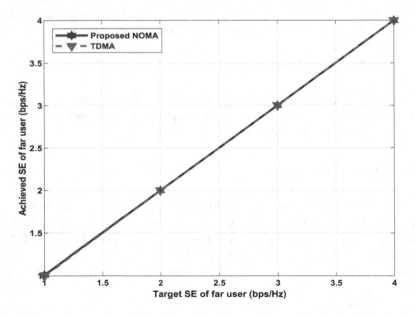

Fig. 3. Achieved SE of far user against target SE

We also evaluate the performance of the proposed approach along with TDMA in terms of power requirement. For that, achieved SE should be same for both the cases. So, target SE of TDMA is kept as the achieved SE of proposed NOMA.

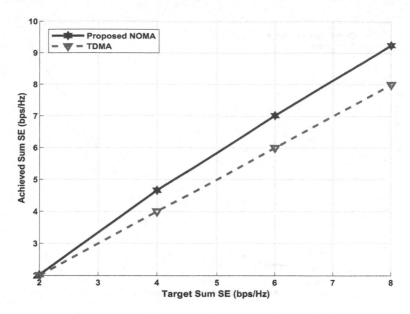

Fig. 4. Achieved sum SE against target sum SE

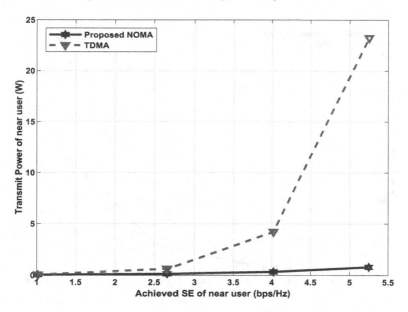

Fig. 5. Required near user's transmit power against achieved SE

The equation for SE of near user NOMA (13) is similar with that of TDMA (16) except the term $\frac{1}{2}$. Hence power required in TDMA i.e. $\|\mathbf{w}_{k-TDMA}\|_2^2$ is more to meet twice the target SE as shown in Fig. 5. But for the far user, the power

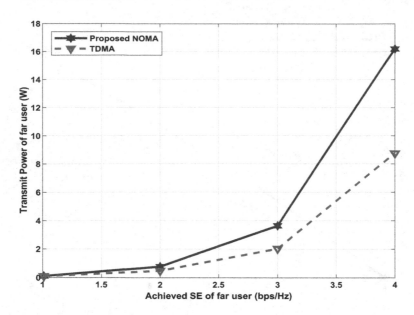

Fig. 6. Required far user's transmit power against achieved SE

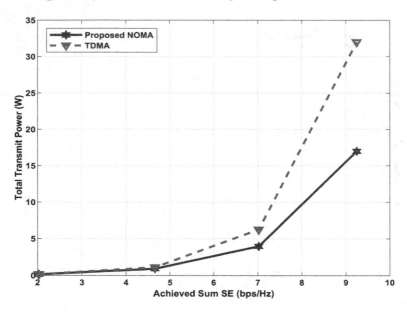

Fig. 7. Required total transmit power against achieved sum SE

requirement in the proposed NOMA approach is more (Fig. 6) to accomplish SIC at near user. From (Fig. 7), it is evident that the proposed approach in NOMA is better compared with the TDMA in terms of the total power requirement for the same achieved sum SE.

5   Conclusion

In this work, we formulated total power minimization problem for two user mm-wave down link NOMA with analog beamforming. SDR method used in this algorithm gives optimum solution since the rank of the PSD matrix involved is always one. Then the weight vector is changed into constant modulus according to the steps discussed in our low complex approach. Moreover, less total power is required in the proposed approach compared with the conventional TDMA to achieve the same sum spectral efficiency. The proposed algorithm can be used in our future work which deals with multi-user NOMA.

References

1. Rappaport, T.S., et al.: Millimeter wave mobile communications for 5G cellular: it will work!. IEEE Access **1**, 335–349 (2013)
2. Kumar, A., Jois, A.S., Ramesh, T.K.: Distance and energy aware device to device communication. In: International Conference on Communication and Signal Processing (ICCSP), pp. 0522–0525 (2019)
3. Ramesh, T.K., Giriraja, C.V.: Study of reassignment strategy in dynamic channel allocation scheme. In: 3rd International Conference on Signal Processing and Integrated Networks, SPIN: Amity School of Engineering and Technology, Noida, India, p. 2016 (2016)
4. Niu, Y., Li, Y., Jin, D., Su, L., Vasilakos, A.V.: A survey of millimeter wave communications (mmWave) for 5G: opportunities and challenges. J. Wirel. Netw. **21**(8), 2657–2676 (2015). https://doi.org/10.1007/s11276-015-0942-z
5. Sheeba Kumari, M., Kumar, N.: Channel model for simultaneous backhaul and access for mmWave 5G outdoor street canyon channel. J. Wirel. Netw. **26**, 5997–6013 (2020). https://doi.org/10.1007/s11276-020-02421-0
6. Zha, M.K., Kumar, N., Lakshmi, Y.V.S.: Performance of zero biased NOMA VLC system. In: 2020 IEEE third 5G World Forum (5GWF), pp. 519–523 (2020)
7. Benjebbour, A., Saito, Y., Kishiyama, Y., Li, A., Harada, A., Nakamura, T.: Concept and practical considerations of non-orthogonal multiple access (NOMA) for future radio access. In: International Symposium on Intelligent Signal Processing and Communication Systems, pp. 770–774 (2013)
8. Riazul Islam, S.M., Avazov, N., Dobre, O.A., Kwak, K.-S.: Power-domain non-orthogonal multiple access (NOMA) in 5G systems: potentials and challenges. IEEE Commun. Surv. Tutor. **19**(2), 721–742 (2017)
9. Andrews, J.G., Bai, T., Kulkarni, M.N., Alkhateeb, A., Gupta, A.K., Heath, R.W.: Modeling and analyzing millimeter wave cellular systems. IEEE Trans. Commun. **65**(1), 403–430 (2017)
10. Sun, S., Rappaport, T.S., Heath, R.W., Nix, A., Rangan, S.: MIMO for millimeter-wave wireless communications: beamforming, spatial multiplexing, or both? IEEE Commun. Mag. **52**(12), 110–121 (2014)
11. Alavi, F., et al.: Beamforming techniques for non-orthogonal multiple access in 5G cellular networks. IEEE Trans. Veh. Technol. **67**(10), 9474–9487 (2018)
12. Chen, Z., Ding, Z., Xu, P., Dai, X.: Comment on Optimal precoding for a QoS optimization problem in two-user MISO-NOMA downlink. IEEE Commun. Lett. **21**(9), 2109–2111 (2017)

13. Huang, Y., Palomar, D.P.: Rank-constrained separable semidefinite programming with applications to optimal beamforming. IEEE Trans. Signal Process. **58**(2), 664–678 (2010)
14. Luo, Z.Q., Ma, W.K., So, A.M.C., Ye, Y., Zhang, S.: Semidefinite relaxation of quadratic optimization problems. IEEE Signal Process. Mag. **27**(3), 20–34 (2010)
15. Ding, Z., Fan, P., Poor, H.V.: Random beamforming in millimeter-wave NOMA networks. IEEE Access **5**, 7667–7681 (2017)
16. Cui, J., Liu, Y., Ding, Z., Fan, P., Nallanathan, A.: Optimal user scheduling and power allocation for millimeter wave NOMA systems. IEEE Trans. Wirel. Commun. **17**(3), 1502–1517 (2018)
17. Xiao, Z., Zhu, L., Choi, J., Xia, P., Xia, X.-G.: Joint power allocation and beamforming for non-orthogonal multiple access (NOMA) in 5G millimeter wave communications. IEEE Trans. Wirel. Commun. **17**(5), 2961–2974 (2018)
18. Sumathi, S., Arpita, T.: Impact of imperfect channel state information on down link sum rate of two user mmWave non orthogonal multiple access. In: IEEE Fourth International Conference on Communication and Electronics Systems (ICCES 2019), pp. 2128–2133 (2019)
19. Kumari, M., Kumar, N., Ramjee, P.: Optimization of street canyon outdoor channel deployment geometry for mmWave 5G communication. AEU - Int. J. Electron. Commun. 153368. https://doi.org/10.1016/j.aeue.2020.153368
20. Zhao, L., Ng, D.W.K., Yuan, J.: Multi-user precoding and channel estimation for hybrid millimeter wave systems. IEEE J. Sel. Areas Commun. **35**(7), 1576–1590 (2017)
21. Grant, M., Boyd, S.: CVX Matlab software for disciplined convex programming, Version 2.0 Beta (2013). http://cvxr.com/cvx

Quantum Communication, IoT and Emerging Applications

A Novel Multi-User Quantum Communication System Using CDMA and Quantum Fourier Transform

M. Anand$^{(\boxtimes)}$ and Pawan Tej Kolusu

Centre for Development of Telematics, Bengaluru 560100, KA, India
{anand.m,pawantej_kolusu}@ieee.org

Abstract. Quantum Communication is an upcoming technology for the next generation communication network. Quantum Communication involves transmitting Quantum Bits (Qubits) instead of regular binary bits in the communication network. With absolute security guaranteed by laws of quantum physics, harnessing this emerging technology is necessary for securing future communication networks. In Quantum Communication, enabling Multi-User communication is needed to support multiple network topologies. There are many open challenges in Multi-User Quantum Communication like security between the transmitter or receiver and router, increasing the difficulty for eavesdroppers to guess the Qubit being exchanged across routers and enabling hierarchical network topology etc. In this paper, the idea is to use Code Division Multiple Access (CDMA) to ensure security between transmitter and transmit side router and also between receiver side router and receiver. The issue of eavesdroppers is solved by using Quantum Fourier Transform (QFT) which transforms incoming Qubits thereby securing them from eavesdroppers. QFT and the corresponding Inverse Quantum Fourier Transform (IQFT) makes the Qubit more secure in the network. QFT and IQFT is also scalable making it ideal for hierarchical network topology. This paper provides the mathematical proof of the security and scalability of the proposed Multi-User Quantum Communication System.

Keywords: Quantum computing · Multi-user quantum communication · Quantum circuits · Quantum internet · CDMA · Quantum Fourier transform

1 Introduction

1.1 Qubit and Quantum Communication

In quantum computing, a qubit or quantum bit is the basic unit of quantum information [1]. Qubit is the quantum version of the classical binary bit physically realized with a two-state device like the spin of the electron in which the two

© ICST Institute for Computer Sciences, Social Informatics and Telecommunications Engineering 2021
Published by Springer Nature Switzerland AG 2021. All Rights Reserved
N. Kumar et al. (Eds.): UBICNET 2021, LNICST 383, pp. 79–90, 2021.
https://doi.org/10.1007/978-3-030-79276-3_7

levels can be taken as spin up and spin down; or the polarization of a single photon in which the two states can be taken to be the vertical polarization and the horizontal polarization.

In a classical system, a bit would have to be in one state or the other. However, quantum mechanics allows the qubit to be in a coherent superposition of both states simultaneously, a property which is fundamental to quantum mechanics and quantum computing [2].

Quantum communication is a field of applied quantum physics closely related to quantum information processing and quantum teleportation. Its most interesting application is protecting information channels against eavesdropping by means of quantum cryptography.

Similar to the way classical networks exchange communications and data between different inter-connected entities, a quantum network enables the secure transmission and exchange of quantum communications (quantum cryptographic keys) over fiber optic cable between distinct, physically-separated quantum processors, or endpoints.

1.2   Multi-User Quantum Communication

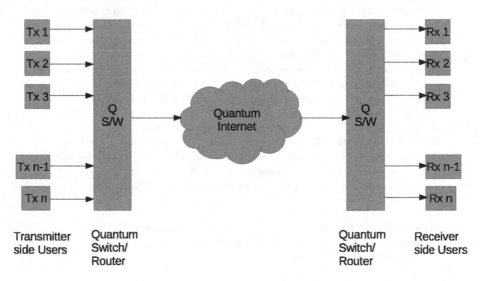

Fig. 1. A multi-user involved quantum communication network system for N transmitter users and N receiver users.

A multi-user involved quantum communication network system for N transmitter users and N receiver users. Routers and switches are used to create a hierarchical network within the quantum internet as shown in Fig. 1. Security of all data flowing across users needs to be ensured end-to-end by all nodes involved. There are many publications in this domain [5–9].

Challenges in Multi-User Quantum Communication:

Security between the transmitter or receiver and router. In the quantum internet each node which is transmitter must encode the data in Qubits which the untrusted Switch or Router cannot decode. This is the primary level of security to be ensured by the transmitter. Similarly the receiver must ensure that the qubits received is as expected and no tampering is done to them. This challenge is particularly high when the transmitter is not connected in a peer-to-peer fashion.

Increasing the difficulty for eavesdropper to guess the Qubit being exchanged across routers. The Switch or router at the transmitter and receiver side must be designed in a fashion to ensure that any eavesdropper in the quantum internet cannot decode the data being transmitted. This node is typically not co-located with the receiver or transmitter. Rather the routers and switches are in the custody of untrusted 3rd parties. This increases the need of securing the Qubits across these routers and switches. Proper algorithms and quantum circuits are needed to Increase the difficulty for eavesdroppers to guess the Qubit being exchanged across routers.

Enabling hierarchical network topology. In a multi-user quantum communication system, there is a need for a hierarchical network topology for a scalable quantum internet. This scaling is possible only when the switches and routers are designed using scalable algorithms and circuits.

In this paper the author creates quantum circuits for the transmit side, receive side and the quantum switch. The technologies used are using the basic building blocks of CDMA technology and Quantum Fourier Transform (QFT) and Inverse Quantum Fourier Transform (IQFT) technique, thereby enabling multi-user quantum communication using CDMA and QFT.

The paper is arranged as follows. In Sect. 2 the proposed multi-user quantum communication system model using QFT is described. This section includes the transmitter side modulation and receiver side demodulation. Mathematical treatment is provided in Sect. 3. Section 4 lists the advantages of the proposed system. Results and discussion are in Sect. 5. Section 6 provides the conclusion and future works planned for this proposed system. Section 7 provides acknowledgment followed by references used in this paper.

2   Proposed Multi-User Quantum Communication System Using QFT

Explanation for the Fig. 2 is as follows

User data are sent in Qubits D_1 and D_2 for user 1 and user 2 respectively. CDMA codes for User 1 is represented by 2 Qubits, together marked as C_1. This is used for doing CDMA encoding and decoding. CDMA codes for User 2 is represented by 2 Qubits, together marked as C_2. This is used for doing CDMA encoding and decoding. The first 2 dotted box on the left side in the image represent the quantum circuit inside transmitter side user 1 and user 2 respectively. This performs the CDMA endoing procedure. The next 2 dotted box in the middle of

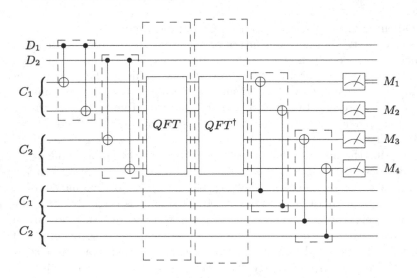

Fig. 2. Proposed Multi-User Quantum circuit.

the image represent transmitter side and receiver side quantum switch or router respectively. The last 2 dotted box towards the right in the image represent the quantum circuit implemented at receiver side user 1 and user 2 respectively. This performs the CDMA decoding procedure. A small operation on M_1 and M_2 is needed to recover the user 1 data D_1 and similarly, a small operation on M_3 and M_4 is needed to recover the user 2 data D_2. More details and explanations are provided in subsequent sections. All quantum circuits in this paper is drawn using qcircuit latex package [10].

2.1 Transmitter Side Modulation

Code Division Multiple Access (CDMA) Encoding

Each classical bit d is encoded as a single Qubit D.

Classical $0 -> |0\rangle$ and Classical $1 -> |1\rangle$

For demonstration purpose this paper will consider a 2-User system henceforth. Let d_1 and d_2 be the classical bits of User1 and User2 respectively. Let D_1 and D_2 be the corresponding Qubits of d_1 and d_2.

Each transmit User is assigned a 2-Qubit Walsh hadamard code-word C.

Let C_1 and C_2 be the walsh hadamard codes assigned to each User.

$$U_1 : C_1 = |C_{1x}C_{1y}\rangle = |01\rangle \tag{1}$$

$$U_2 : C_2 = |C_{2x}C_{2y}\rangle = |00\rangle \tag{2}$$

The CDMA encoding includes a XOR operation using a C−NOT gate. Each User Qubit D is used as a Control bit for doing XOR operation on the respective User code-word C as shown in Fig. 3.

Fig. 3. CDMA encoding for single User

The CDMA encoded Qubits for individual Users is computed as shown in Fig. 4

$$E_1 = |E_{1x}E_{1y}\rangle \, where, \, E_{1x} = C_{1x} \oplus D_1 \, and \, E_{1y} = C_{1y} \oplus D_1 \qquad (3)$$

$$E_2 = |E_{2x}E_{2y}\rangle \, where, \, E_{2x} = C_{2x} \oplus D_2 \, and \, E_{2y} = C_{2y} \oplus D_2 \qquad (4)$$

Using equations (1),(2), (3) and (4) we get

$$E_1 = |D_1 D_1'\rangle \qquad (5)$$

$$E_2 = |D_2 D_2\rangle \qquad (6)$$

The CDMA encoded data for both Users is given by (7)

$$E = |E_{1x}E_{1y}E_{2x}E_{2y}\rangle = |D_1 D_1' D_2 D_2\rangle \qquad (7)$$

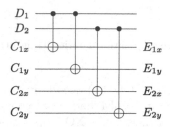

Fig. 4. CDMA encoding for 2 User system

Quantum Fourier Transform (QFT)
Quantum fourier transform is used as a channel encoding method to modulate the CDMA encoded data. For a n User system we use $2n$-Qubit QFT model. For example for a 2 User system, we use 4 Qubit QFT model. The CDMA encoded data E is passed through the QFT block as shown in Fig. 5.

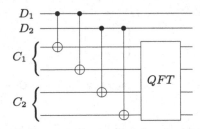

Fig. 5. QFT of CDMA encoded data of 2 Users

2.2   Receiver Side Demodulation

Inverse Quantum Fourier Transform (IQFT)
    The received signal is passed through the IQFT block as shown in Fig. 6. The initially CDMA encoded signal $E = |D_1 D_1' D_2 D_2\rangle$ is recovered after this block.

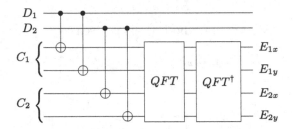

Fig. 6. Inverse Quantum Fourier Transform circuit

Code Division Multiple Access (CDMA) decoding
Each receiver will have the same walsh hadamard codes C_1 and C_2 respectively. The CDMA decoding part consists of a XOR operation using a C–NOT gate. Each User code Qubits are used as Control bits for doing XOR operation on the IQFT decoded Qubits as shown in Fig. 7.
    Let $F = |F_{1x}F_{1y}F_{2x}F_{2y}\rangle$ be the CDMA decoded Qubits, then as per Fig. 7

$$F_{1x} = C_{1x} \oplus E_{1x};\ F_{1y} = C_{1y} \oplus E_{1y};\ F_{2x} = C_{2x} \oplus E_{2x}\ and\ F_{2y} = C_{2y} \oplus E_{2y} \quad (8)$$

Measurement and User bit recovery
As a next step, we measure all the CDMA decoded Qubits F_{1x}, F_{1y}, F_{2x} and F_{2y}. Let M_1, M_2, M_3 and M_4 be the measured values respectively. The User classical data bits can be recovered by performing an Logical AND operation on the measured values.

$$d_1 = M_1 \wedge M_2\ and\ d_2 = M_3 \wedge M_4 \quad (9)$$

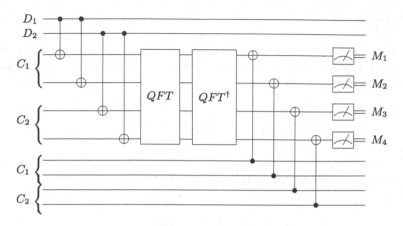

Fig. 7. Quantum circuit for CDMA Decoding

3   Mathematical Example for Single User System

Considering a single User system, where the transmit User classical bit is $d_1 = $
'1'. Converting the classical bit to Qubit we get $D_1 = |1\rangle$. In the real world the $|0\rangle$
may be represent on spin-up of electron and $|1\rangle$ may be represent on spin-down
of electron. Alternatively $|0\rangle$ may be represent on vertical polarization of photon
and $|1\rangle$ may be represent horizontal polarization of photon.

3.1   CDMA Encoding

Let $C_1 = |C_{1x}C_{1y}\rangle = |01\rangle$ be the walsh hadamard code assigned to User1.
    The CDMA encoded Qubits for User1 is computed as

$$E_1 = |E_{1x}E_{1y}\rangle \tag{10}$$

where,

$$E_{1x} = C_{1x} \oplus D_1 = |0\rangle \oplus |1\rangle = |1\rangle \tag{11}$$

and

$$E_{1y} = C_{1y} \oplus D_1 = |1\rangle \oplus |1\rangle = |0\rangle \tag{12}$$

$$Hence\, E_1 = |10\rangle \tag{13}$$

$$\text{Vector representation of } E_1 = \begin{bmatrix} 0 \\ 0 \\ 1 \\ 0 \end{bmatrix}$$

3.2 QFT

Since the mathematical example is for 1 user, 2 Qubit QFT is needed. The QFT matrix for 2 qubit system is represented by QFT_4 which is represented in matrix form as shown in equation below:

$$QFT_4 = 1/2 \begin{bmatrix} 1 & 1 & 1 & 1 \\ 1 & i & -1 & -i \\ 1 & -1 & 1 & -1 \\ 1 & -i & -1 & i \end{bmatrix} \tag{14}$$

The operation of QFT over the output of CDMA encoded Data E_1 is represented by the equation below:

$$QFT * E_1 = 1/2 \begin{bmatrix} 1 & 1 & 1 & 1 \\ 1 & i & -1 & -i \\ 1 & -1 & 1 & -1 \\ 1 & -i & -1 & i \end{bmatrix} \cdot \begin{bmatrix} 0 \\ 0 \\ 1 \\ 0 \end{bmatrix} = 1/2 \begin{bmatrix} 1 \\ -1 \\ 1 \\ -1 \end{bmatrix} = G_1 \tag{15}$$

3.3 IQFT

Since the mathematical example is for 1 user, 2 Qubit IQFT is needed. The TQFT matrix for 2 qubit system is represented by $IQFT_4$ which is represented in matrix form as shown in equation below:

$$IQFT_4 = 1/2 \begin{bmatrix} 1 & 1 & 1 & 1 \\ 1 & -i & -1 & i \\ 1 & -1 & 1 & -1 \\ 1 & i & -1 & -i \end{bmatrix} \tag{16}$$

The operation of IQFT over the output of QFT modulated Data G_1 is represented by the equation below:

$$IQFT * G_1 = 1/2 \begin{bmatrix} 1 & 1 & 1 & 1 \\ 1 & -i & -1 & i \\ 1 & -1 & 1 & -1 \\ 1 & i & -1 & -i \end{bmatrix} \cdot 1/2 \begin{bmatrix} 1 \\ -1 \\ 1 \\ -1 \end{bmatrix} = \begin{bmatrix} 0 \\ 0 \\ 1 \\ 0 \end{bmatrix} = E_1 \tag{17}$$

3.4 CDMA Decoding

Ket representation of $E_1 = |10\rangle$

$$F_{1x} = C_{1x} \oplus E_{1x}; = |0\rangle \oplus |1\rangle = |1\rangle \tag{18}$$

$$F_{1y} = C_{1y} \oplus E_{1y}; = |1\rangle \oplus |0\rangle = |1\rangle \tag{19}$$

3.5   Measurement and User Bit Recovery

After Measuring we have $M_1 = 1\, and\, M_2 = 1$.

$$d_1 = M_1 \wedge M_2 = 1 \wedge 1 = 1. \qquad (20)$$

4   Simulation of 2 User System

The proposed quantum communication network was simulated using Quirk quantum simulator tool [10] for 2 transmit users and 2 receive users.

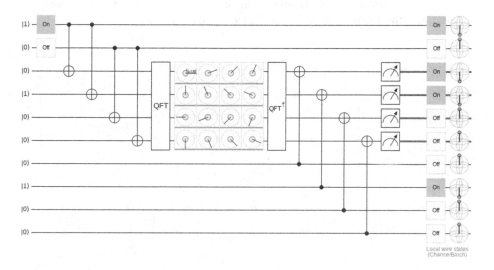

Fig. 8. Quantum communication network simulation for 2 transmit users and 2 receive users with user data bits $d_1 = 1$ and $d_2 = 0$.

In this simulation shown in Fig. 8. The data for user 1 is shown in the first line. The data for user 2 is shown in the 2nd line. In this simulation shown in figure the user 1 want to transmit binary data 1, which is mapped to quantum bit (qubit) $|1\rangle$. The measurement of this qubit would be high or on state represented by the box which shows "On". This box is a hypothetical measurement result. Similarly user 2 want to transmit binary data 0, which is mapped to quantum bit (qubit) $|0\rangle$. The measurement of this qubit would be low or off state represented by the box which shows "Off". This box is a hypothetical measurement result.

The next 2 lines are the Code for user 1. CDMA code for user 1 is $|10\rangle$ which is represented in the 2 lines with qubit values $|1\rangle$ and $|0\rangle$. Similarly the next 2 lines are the Code for user 2. CDMA code for user 2 is $|00\rangle$ which is represented in the 2 lines with qubit values $|0\rangle$ and $|0\rangle$. The CDMA encoding is done using C-NOT gates. This operation is represented by first 2 vertical lines drop shown in figure. The output of this operation is the CDMA encoded data. Similarly the next 2 vertical lines drop shown in figure are CDMA encoding for user 2.

These 4 encoded outputs are fed to the QFT module at the switch. The Output of the QFT is transmitted to the receiver side router or switch over the quantum internet. The simulation shows the spreading of data. It is very difficult to guess or decode the user data. At the receiver side router or switch, Inverse QFT is performed and the first 2 qubits are sent to User 1 at receiver and the next 2 Qubit to user 2 at receiver side. At user 1 at receiver, C-NOT operation is again performed using the same code as done during the encoding phase. The resulting Qubit is measured. The 2 measured outputs are in binary bits. Binary logic "AND" gate operation is performed on these 2 bits. For user 1 the values are "on" which means 1. AND operation of two 1's is 1 which is the user 1's binary data. Similar operation is performed by user 2.

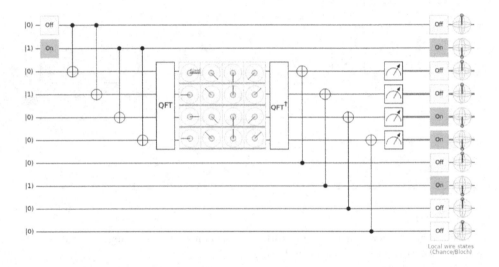

Fig. 9. Quantum communication network simulation for 2 transmit users and 2 receive users with user data bits $d_1 = 0$ and $d_2 = 1$.

Similar simulation is done with user 1 binary data 0 and user 2 binary data 1 using the same CSMA codes as shown in Fig. 9. The result of measurement for user 1 produces 2 "off" sates representing 0 and 0. The data for user 1 is retrieved using binary "AND" operation which results in 0 meaning user 1 data is 0. Similarly the result of measurement for user 2 produces 2 "on" sates representing 1 and 1. The data for user 2 is retrieved using binary "AND" operation which results in 1 meaning user 2 data is 1.

These simulation provides proof that the proposed multi-user quantum communication system using QFT is robust, secure and scalable.

5   Advantages of the Proposed System

The following are the advantages of the proposed system.

1. Increased Security between the transmitter or receiver and router using CDMA codes. Using simple C-NOT gates at the transmitter and receiver side has greatly increased the security between end-users and routers.
2. Increasing the difficulty for eavesdropper to guess the Qubit being exchanged across routers by using Quantum Fourier Transform. Using Quantum Fourier Transform which transforms the Qubit from one-base to another greatly increases security by spreading the each user information across multiple new basis vectors. This QFT is typically carried out in Quantum computers. The eavesdropper must also have an equal or more stronger Quantum Computer to get the CDMA encoded data. Still the user data is secure from eavesdropping.
3. Enabling hierarchical network topology by using QFT scaling. QFT can be scaled linearly from N to N+1 Qubits order unlike classical DFT where scaling happens on powers of 2.

6   Results and Discussion

In this paper a novel Multi-user Quantum Communication system using CDMA and QFT has been proposed. The proposed system uses qubits to exchange data between transmitter and receiver in the Quantum internet using quantum switches or routers.

The proposed network architecture provides security and scalability. The security between the transmitter or receiver and router is by using CDMA codes and the mathematical proof has been provided for the same.

The need for increasing the difficulty for eavesdroppers to guess the qubit being exchanged across routers done by using Quantum Fourier Transform and its inverse IQFT. The mathematical treatment for the same has been provided in this paper.

The scalability is also derived from the fact that QFT can be scaled linearly. The proposed system is robust and can be implemented with minimal requirement of C-NOT gate at the user side and QFT at the router or switches thereby making it an elegant Multi-User Quantum Communication System.

7   Conclusion and Future Works

In conclusion, the paper has provided a new out-look on the quantum internet for a multi-user scalable network. The physical realization is still some time away and there are many new on-going methods to optimize the QFT system. The future works would include optimizing the overall system on the number of quantum gates and simplifying the QFT system.

Acknowledgment. The authors would like to acknowledge the time resource provided by Centre for Development of Telematics, Bengaluru, India.

References

1. Nielsen, M.A., Chuang, I.L.: Quantum Computation and Quantum Information, 2nd edn. Cambridge University Press/Massachusetts Institute of Technology, Cambridge (2000)
2. Qubit Wikipedia. https://en.wikipedia.org/wiki/Qubit. Accessed 9 Oct 2020
3. Quantiki. https://www.quantiki.org/wiki/quantum-gates. Accessed 9 Oct 2020
4. Science Direct. https://www.sciencedirect.com/topics/computer-science/quantum-circuit. Accessed 9 Oct 2020
5. Sharma, V., Banerjee, S.: Quantum communication using code division multiple access network. Opt. Quant. Electron. **52**, 381 (2020). https://doi.org/10.1007/s11082-020-02494-3
6. Tan, X., Cheng, S., Li, J., Feng, Z.: Quantum key distribution protocol using quantum fourier transform. In: 2015 IEEE 29th International Conference on Advanced Information Networking and Applications Workshops, Gwangiu, pp. 96–101 (2015). https://doi.org/10.1109/WAINA.2015.8
7. Kumavor, P.D., Beal, A.C., Yelin, S., Donkor, E., Wang, B.C.: Comparison of four multi-user quantum key distribution schemes over passive optical networks. J. Lightwave Technol. **23**(1), 268–276 (2005). https://doi.org/10.1109/JLT.2004.834481
8. Brassard, G., Bussieres, F., Godbout, N., Lacroix, S.: Multiuser quantum key distribution using wavelength division multiplexing. In: Proceedings of SPIE 5260, Applications of Photonic Technology 6, (2003). https://doi.org/10.1117/12.543338
9. Xue, P., Wang, K., Wang, X.: Efficient multiuser quantum cryptography network based on entanglement. Sci. Rep. **7**, 45928 (2017). https://doi.org/10.1038/srep45928
10. Quirk - A drag-and-drop quantum circuit simulator. https://algassert.com/quirk. Accessed 9 Oct 2020
11. qcircuit - Macros to generate quantum ciruits. https://ctan.org/pkg/qcircuit. Accessed 9 Oct 2020

Qubit Share Multiple Access Scheme (QSMA)

Pawan Tej Kolusu[✉] and M. Anand

Centre for Development of Telematics, Bengaluru 560100, KA, India
{pawantej_kolusu,anand.m}@ieee.org

Abstract. Quantum computing has shown great advancement in recent times. With significant properties like Quantum superposition and Quantum entanglement, the time required to evaluate a function in polynomial time has reduced significantly. The meta stable nature of the Quantum bits (Qubits), has opened doors for a wide research in network optimization and security. This paper proposes a novel Multi-user Qubit sharing scheme called the QSMA. Classical bits of information can be shared among multiple transmit and receive users using minimal number of Qubits. Protocols like superdense coding have been used to encode classical bits of information to Qubits. Mathematical transformation tools like the Quantum Fourier transform (QFT) have been incorporated to enhance the security of the QSMA system.

Keywords: Quantum computing · Multi-user quantum communication · Quantum circuits · Quantum internet · QSMA · Quantum Fourier transform · Superdense coding

1 Introduction

Quantum computation, derived from Quantum physics, is an emerging field which has shown significant progress and applications in the field of networking and security. Its inbuilt properties like superposition and entanglement have made way for Quantum computers which can solve polynomial equations much faster than a classical computer. The other important application called the Quantum key distribution is widely used for security purpose.

1.1 Qubits and Measurement

Quantum computing is based on the Quantum bits or Qubits. They are a linear combination of computational basis states. The spin of an electron or a photon polarization can be considered as examples of Qubit. Mathematically, a Qubit is defined as $|\psi\rangle = \alpha|0\rangle + \beta|1\rangle$, where $|0\rangle$ and $|1\rangle$ are orthogonal computational basis states. α and β are complex numbers. The Qubit exists simultaneously in $|0\rangle$ and $|1\rangle$ states which is called the 'superposition'. This is said to be in a quasi

© ICST Institute for Computer Sciences, Social Informatics and Telecommunications Engineering 2021
Published by Springer Nature Switzerland AG 2021. All Rights Reserved
N. Kumar et al. (Eds.): UBICNET 2021, LNICST 383, pp. 91–104, 2021.
https://doi.org/10.1007/978-3-030-79276-3_8

stable state and when 'Measured', gives a stable state $|0\rangle$ or $|1\rangle$ with probability $|\alpha|^2$ or $|\beta|^2$ respectively. Also,

$$|\alpha|^2 + |\beta|^2 = 1 \tag{1}$$

Alternately, a qubit can be written in vector form as

$$\begin{bmatrix} \alpha \\ \beta \end{bmatrix} \tag{2}$$

Multiple Qubits can be represented by tensor product of individual Qubits. A system with n-Qubits can be represented as

$$|x_0 x_1 x_2 ... x_{n-1}\rangle = \sum_{k=0}^{n-1} \alpha_k |x_k\rangle \tag{3}$$

where,

$$\sum_{k=0}^{n-1} |\alpha_k|^2 = 1 \tag{4}$$

In vector form qubits can be written as

$$\begin{bmatrix} \alpha_0 \\ \alpha_1 \\ ... \\ ... \\ \alpha_{n-1} \end{bmatrix} \tag{5}$$

1.2   Quantum Gates and Circuits

Just like classical gates, quantum computing uses quantum gates. For our reference, we have used the following quantum gates, Identity, Hadamard, Pauli-X, Pauli-Z, Controlled-NOT and Controlled-Z gates. The matrix form of the gates can be found in [1].

1.3   Quantum Fourier Transform (QFT)

In quantum computing, the quantum Fourier transform (QFT) is a linear transformation on quantum bits, and is the quantum analogue of the inverse discrete Fourier transform.

The quantum Fourier transform can be performed efficiently on a quantum computer.

In case that $|x\rangle$ is a basis state, the quantum Fourier Transform can be represented as

$$\frac{1}{\sqrt{N}} \sum_{k=0}^{N-1} \omega_N^{xk} |k\rangle \tag{6}$$

In case N = 4, the transformation matrix becomes

$$F_4 = 1/2 \begin{bmatrix} 1 & 1 & 1 & 1 \\ 1 & i & -1 & -i \\ 1 & -1 & 1 & -1 \\ 1 & -i & -1 & i \end{bmatrix} \tag{7}$$

1.4 Inverse Quantum Fourier Transform (IQFT)

The inverse quantum Fourier transform (IQFT) is a linear transformation on quantum bits, and is the quantum analogue of the discrete Fourier transform.

In case that $|k\rangle$ is a basis state, the inverse quantum Fourier Transform can represented as

$$\frac{1}{\sqrt{N}} \sum_{k=0}^{N-1} \omega_N^{-nk} |x\rangle \tag{8}$$

In case N = 4, the transformation matrix becomes

$$IQFT_4 = 1/2 \begin{bmatrix} 1 & 1 & 1 & 1 \\ 1 & -i & -1 & i \\ 1 & -1 & 1 & -1 \\ 1 & i & -1 & -i \end{bmatrix} \tag{9}$$

1.5 Challenges in Multi-user Quantum Communication and Quantum Internet

Security: The classical bits which are encoded to qubits must be secure throughout the communication channel away from the risk of decoding by an evesdropper. The main challenge is to ensure security at each intermediate node. Hence the security algorithms used to encode must be complex and random in nature. The idea is to transform the qubits over a computationally large eigen basis.

Scalability: The multi-user quantum communication system must be flexible for upgradation to higher order system using minimal changes.

Optimization: The multi-user quantum communication system involves large number of mathematical transformations. Hence more number of gates are required to evaluate when compared to a classical communication system.

1.6 The Paper is Arranged as Follows

The proposed Multi-User Quantum Communication System Model using QSMA is illustrated in Sect. 2. This section includes basic architecture for a n-user system using QSMA. QSMA for 2-user system is described in Sect. 3. Mathematical model for2-user system is provided in Sect. 4. Simulations are present in Sect. 5. Section 6 lists the advantages of the proposed system. Results and Discussion are in Sect. 7. Section 8 provides the conclusion and future works planned for this proposed system. The last section provides acknowledgment followed by references used in this paper.

2   Multi-user Quantum Communication Using QSMA

A multi-user quantum communication network system for 2K transmitter users and 2K receiver users is proposed. Figure 1 shows K transmitter side users sending Classical Binary Bits to QSMA Transmitter 1 which converts these binary bit sequence to Qubit using Super Dense Coding. The output of these are sent to transmitter side switch which does QFT operation. At the receiver side the IQFT is performed by Receiver side Switch and send the recovered qubits to QSMA receivers, which in turn does Superdense decoding and send binary bits to respective users.

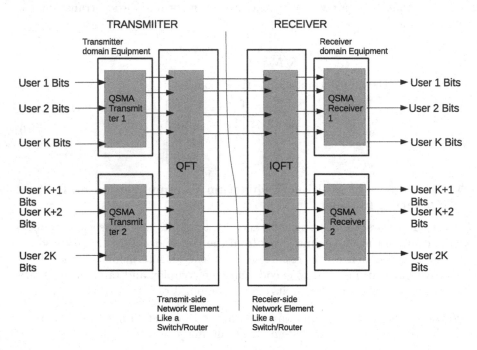

Fig. 1. Multi user QSMA system

2.1   Qubit Share Multiple Access Scheme (QSMA)

The idea of QSMA is to multiplex the classical user bits using the proposed QSMA circuit. The QSMA circuit contains multiple Superdense coding circuits based on the number of Users. For example, let us consider the case where 4 classical data bits is considered. We need 2 superdense coding circuits to implement the same. Our Qubit sharing idea starts with a multiplexing unit say M embedded in our QSMA circuit. Our theory is based on the fact that these four data bits can come from different number of cases/Users as shown below

1. Two users with two classical bits each.
2. Three Users, with U1- 1bit, U2- 1 bit and U3- 2 bits.
3. Four Users with each one classical bit.

As we can see, all the above 3 cases can be accommodated using our proposed QSMA circuit for 4 classical bits. Our following sections describe in detail about case-1.

2.2 Super Dense Coding

The superdense coding can be divided into three steps as given below [4].

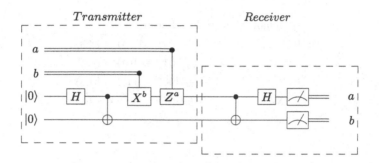

Fig. 2. Superdense coding circuit

Step1: Entangled Bell Pair. The Superdense coding starts with a third party say Charlie which has two Qubits. These two Qubits are processed to form an entangled Bell pair. The step by step procedure to form an entangled Bell pair state is given below. Both the Qubits are initially set to $|00\rangle$. Then, a Hadamard gate (H) is applied on the first Qubit $|0\rangle$ to create the superposition $|+\rangle$. So we get the state as,

$$|+0\rangle = \frac{1}{\sqrt{2}}(|00\rangle + |10\rangle) \tag{10}$$

After that, a CNOT gate (CX) is applied using the first Qubit $|+\rangle$ as a control and the second Qubit $|0\rangle$ as the target. Here we get the entangled bell pair state as

$$\frac{1}{\sqrt{2}}(|00\rangle + |11\rangle) \tag{11}$$

Step2: Superdense Encoding. Charlie sends the entangled Bell pair Qubits to two people, say Alice and BOB. Alice receives the first Qubit and BOB receives the second Qubit. The main idea of this protocol is for Alice to send 2 classical bits of information to Bob using her qubit. But before Alice sends, she needs to apply a set of quantum gates to her qubit depending on the 2 bits of classical information she wants to send to BOB. This is achieved using a controlled Z and a controlled NOT gate in sequence as shown in Fig. 2.

The Table 1 shows the Quantum gates required for encoding each pair of classical bits:

There are 4 cases based on two-bit strings a and b:

Table 1. Superdense encoding table

a	b	Quantum gate	Final state
0	0	I	$\frac{1}{\sqrt{2}}(|00\rangle + |11\rangle) = |\beta_{00}\rangle$
0	1	X	$\frac{1}{\sqrt{2}}(|01\rangle + |10\rangle) = |\beta_{01}\rangle$
1	0	Z	$\frac{1}{\sqrt{2}}(|00\rangle - |11\rangle) = |\beta_{10}\rangle$
1	1	XZ	$\frac{1}{\sqrt{2}}(|01\rangle - |10\rangle) = |\beta_{11}\rangle$

Step3: Superdense Decoding. Alice then sends its encoded qubit to BOB and BOB uses his qubit to decode Alice's message. Bob applies a CNOT gate using the Alice's qubit as control and Charlie's Qubit as target. BOB then applies a Hadamard gate and performs a measurement on both qubits to extract Alice's message. The step by step decoding process is shown in Table 2.

Table 2. Superdense decoding table

Initial state	After CNOT	After H	a	b
$\frac{1}{\sqrt{2}}(|00\rangle + |11\rangle)$	$\frac{1}{\sqrt{2}}(|00\rangle + |10\rangle)$	$|00\rangle$	0	0
$\frac{1}{\sqrt{2}}(|01\rangle + |10\rangle)$	$\frac{1}{\sqrt{2}}(|01\rangle + |11\rangle)$	$|01\rangle$	0	1
$\frac{1}{\sqrt{2}}(|00\rangle - |11\rangle)$	$\frac{1}{\sqrt{2}}(|00\rangle - |10\rangle)$	$|10\rangle$	1	0
$\frac{1}{\sqrt{2}}(|01\rangle - |10\rangle)$	$\frac{1}{\sqrt{2}}(|01\rangle - |11\rangle)$	$|11\rangle$	1	1

3   Proposed Qubit Share Multiple Access Scheme for 2-Users

3.1   Transmitter Side Modulation

For demonstration purpose we consider a 2-User system with two classical bits each. Let a_1 and b_1 be the classical bits of User1 and a_2 and b_2 be the classical bits of User2.

Superdense Encoding. Let the 3rd party provide four Qubits $|Q_1Q_2Q_3Q_4\rangle$ for encoding the four classical data bits using super dense coding. For a n User system we use n-Superdense coding circuits. Initially

$$|Q_1Q_2Q_3Q_4\rangle = |0000\rangle \tag{12}$$

We use two superdense encoding circuits for 2-users as shown in Fig. 3. Based on Sect. 2.2, let the classical bits $[a_1b_1]$ of User-1 be encoded using the Qubits $|Q_2Q_1\rangle$ using the quantum gates as specified in Table 1. Similarly let the classical bits $[a_2b_2]$ of User-2 be encoded using the Qubits $|Q_3Q_4\rangle$. Let the system state after superdense encoding be

$$|Q_{1S}Q_{2S}Q_{3S}Q_{4S}\rangle \tag{13}$$

Quantum Fourier Transform (QFT). Quantum fourier transform is used as a channel encoding method to modulate the superdense encoded Qubits. For a n-User system we use n-Qubit QFT model. For example for a 2-User system, we use 2- Qubit QFT model. The superdense encoded Qubits $|Q_{2S}\rangle$ and $|Q_{3S}\rangle$ are passed through the QFT block as shown in Fig. 3. Please note that we send only these two Qubits for QFT encoding as the transformation is required only on these two Qubits. Then, the system state after QFT encoding becomes

$$|Q_{1S}Q_{2SE}Q_{3SE}Q_{4S}\rangle \tag{14}$$

3.2   Receiver Side Demodulation

Inverse Quantum Fourier Transform (IQFT). The received signal is passed through the IQFT block as shown in Fig. 3. After IQFT, the superdense encoded Qubits $|Q_{2s}\rangle$ and $|Q_{3s}\rangle$ are recovered. So, the system state after IQFT decoding becomes

$$|Q_{1S}Q_{2S}Q_{3S}Q_{4S}\rangle \tag{15}$$

3.3   Superdense Decoding

Based on Sect. 2.2, we decode the classical bits of User-1 and User-2. We use two superdense decoding circuits as shown in Fig. 3. Let the system state after superdense decoding be

$$|Q_{1SD}Q_{2SD}Q_{3SD}Q_{4SD}\rangle \tag{16}$$

3.4   Measurement and User Bit Recovery

After measuring the four Qubits $|Q_{1SD}\rangle$, $|Q_{2SD}\rangle$, $|Q_{3SD}\rangle$ and $|Q_{4SD}\rangle$ we get the classical bits b_1, a_1, a_2 and b_2 respectively as shown in Fig. 3.

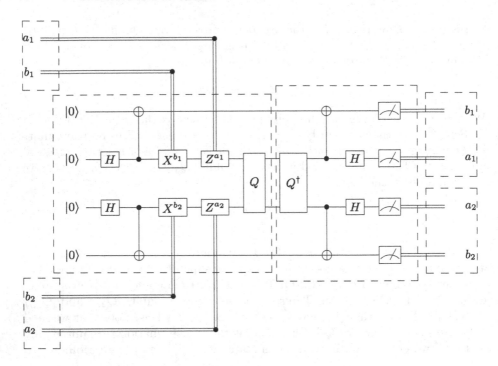

Fig. 3. Proposed QSMA circuit for 2-users

4   Mathematical Example for 2-User System

Considering a 2-User system, where the transmit User-1 classical bits are $[a_1 b_1]$ = [01] and transmit User-2 classical bits are $[a_2 b_2]$ = [10].

4.1   Superdense Encoding

Each User has a superdense coding circuit. Hence for each User, computation can be done separately using two Qubits. The upper section

$$|Q_2 Q_1\rangle \tag{17}$$

is for User-1 and the lower section

$$|Q_3 Q_4\rangle \tag{18}$$

is for User-2.

Note: For Upper section as per Fig. 3, the superdense coding circuit is upside down. Hence the order $|Q_2 Q_1\rangle$ is considered henceforth.

For User-1, the initial state is

$$|Q_2Q_1\rangle = |00\rangle \tag{19}$$

After applying Hadamard gate and C-NOT gate, we get the entangled state as

$$\frac{1}{\sqrt{2}}(|00\rangle + |11\rangle) \tag{20}$$

After applying 'X' gate on $|Q_2\rangle$, we get the state as

$$|Q_{2S}Q_{1S}\rangle = \frac{1}{\sqrt{2}}(|10\rangle + |01\rangle) \tag{21}$$

For User-2, the initial state is

$$|Q_3Q_4\rangle = |00\rangle \tag{22}$$

After applying Hadamard gate and C-NOT gate, we get the entangled state as

$$\frac{1}{\sqrt{2}}(|00\rangle + |11\rangle) \tag{23}$$

After applying 'Z' gate on $|Q_3\rangle$, we get the state as

$$|Q_{3S}Q_{4S}\rangle = \frac{1}{\sqrt{2}}(|00\rangle - |11\rangle) \tag{24}$$

4.2   QFT

2-Qubit QFT is applied to the two superdense encoded Qubits $|Q_{2S}\rangle$ and $|Q_{3S}\rangle$. let the system state after QFT be

$$|Q_{1S}Q_{2SE}Q_{3SE}Q_{4S}\rangle \tag{25}$$

4.3   IQFT

IQFT on the received signal will give the initial superdense coded Qubits because $QFT * IQFT = I$ as shown in Fig. 4 So, then the system state after IQFT will be

$$|Q_{1S}Q_{2S}Q_{3S}Q_{4S}\rangle \tag{26}$$

After IQFT, For User-1, the state is

$$|Q_{2S}Q_{1S}\rangle = \frac{1}{\sqrt{2}}(|10\rangle + |01\rangle) \tag{27}$$

After IQFT, For User-2, the state is

$$|Q_{3S}Q_{4S}\rangle = \frac{1}{\sqrt{2}}(|00\rangle - |11\rangle) \tag{28}$$

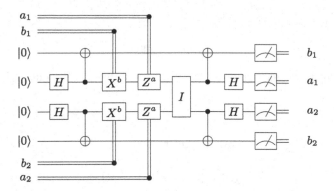

Fig. 4. Indentity property of QFT and IQFT blocks

4.4   Superdense Decoding

For Superdense decoding, first a controlled-NOT is applied
For User-1, the state becomes

$$\frac{1}{\sqrt{2}}(|11\rangle + |01\rangle) \tag{29}$$

For User-2, the state becomes

$$\frac{1}{\sqrt{2}}(|00\rangle - |10\rangle) \tag{30}$$

Then, after applying Hadamard gate,
For User-1, the state becomes

$$|Q_{2SD}Q_{1SD}\rangle = |01\rangle \tag{31}$$

For User-2, the state becomes

$$|Q_{3SD}Q_{4SD}\rangle = |10\rangle \tag{32}$$

4.5   Measurement and User Bit Recovery

After measuring the four Qubits $|Q_{2SD}\rangle$, $|Q_{1SD}\rangle$, $|Q_{3SD}\rangle$ and $|Q_{4SD}\rangle$ we get the classical bits $a_1 =$ '0', $b_1 =$ '1', $a_2 =$ '1' and $b_2 =$ '0' respectively.

5   Simulation of 2 User System

The proposed Quantum communication network was simulated using Quirk Quantum simulator tool [10] for 2 transmit users and 2 receive users.

In this simulation the classical data for user 1 is shown in the top two lines. The classical data for user 2 is shown in the bottom two lines. Lines 3 and 4

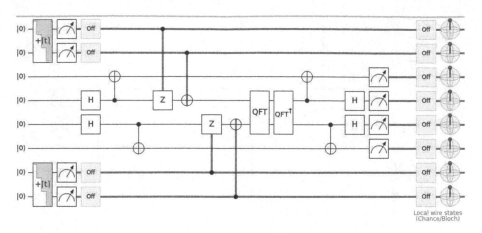

Fig. 5. Quantum communication network simulation for 2 transmit users and 2 receive users with user data bits $a_1 = 0$ and $b_1 = 0$ and $a_2 = 0$ and $b_2 = 0$.

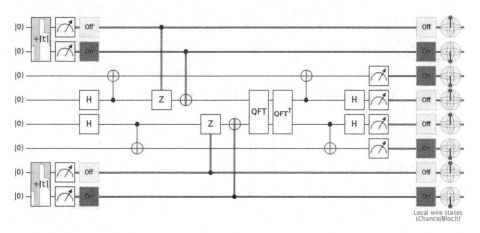

Fig. 6. Quantum communication network simulation for 2 transmit users and 2 receive users with user data bits $a_1 = 0$ and $b_1 = 1$ and $a_2 = 0$ and $b_2 = 1$.

simulate the superdense codingcircuit for user-1 and lines 5 and 6 simulate the superdense coding circuit for user-2. The 2 encoded qubits on one lines 4 and 5 are fed to the QFT module at the switch. The Output of the QFT is transmitted to the receiver side router/switch over the Quantum Internet. At the receiver side router/switch, Inverse QFT is performed and the first 2 qubits on lines 3 and 4 are sent to User 1 at receiver and the next 2 Qubits on lines 5 and 6 are sent to user 2 at receiver side. At the receiver, C-NOT operation followed by Hadamard is performed on lines 3,4,5 and 6. The resulting Qubits on lines 3,4,5 and 6 are measured. The 4 measured outputs are in classical form. As expected, the receiver data is matching with classical data sent. These simulation results show that the proposed multi-user Quantum Communication system using QFT

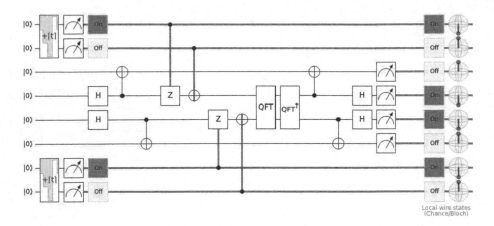

Fig. 7. Quantum communication network simulation for 2 transmit users and 2 receive users with user data bits $a_1 = 1$ and $b_1 = 0$ and $a_2 = 1$ and $b_2 = 0$.

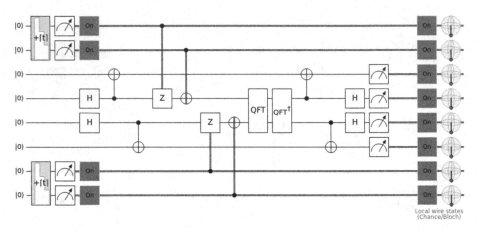

Fig. 8. Quantum communication network simulation for 2 transmit users and 2 receive users with user data bits $a_1 = 1$ and $b_1 = 1$ and $a_2 = 1$ and $b_2 = 1$.

is secured and can be scaled for more number of users. Simulation results for 4 different combinations of classical bits are shown in Fig. 5, Fig. 6, Fig. 7 and Fig. 8.

6 Advantages of the Proposed System

The following are the advantages of the proposed Multi User system.

1. Optimization: Minimal number of Qubit operations are required for implementation. With the help of super dense coding, only one qubit is modified by the User Alice using quantum gates and QFT.

2. *Added security:* The quantum fourier transform (QFT) adds additional security over the standard super dense coding. It becomes highly difficult for the evesdropper to decode the qubits.

3. *Scaling:* The proposed 2-User system can be scaled to a n-User system linearly. One has to only use higher number of superdense coding circuits and higher order QFT and IQFT blocks.

4. *User data multiplexing using QSMA:* The user data bits can be multiplexed in various combinations using the proposed QSMA scheme. This enhances the robustness of the Multi-user system.

7   Results and Discussion

In this paper, a multi user communication system is proposed using QSMA. It is shown that the user classical bits can be transmitted and received using superdense coding. The QFT/ IQFT encoding/decoding at the core network provides extra security making it difficult for the evesdropper to decode the user bits. Simulation results for various combination of classical bits are also shown. The mathematical model for the proposed system is also shown. The QSMA system for higher number of users can be implemented as the proposed system is scalable. One has to only use higher number of superdense coding circuits and higher order QFT and IQFT blocks. Different ways of combining user data using a multiplexing unit is also shown.

8   Conclusion and Future Works

In conclusion, the paper has provided a novel multi-user scalable network for the quantum internet. Though the practical implementation might take time, but owing to the recent advancement in quantum physics, one can be optimistic for realizing a physical system in near future. The future works would include optimizing the total number of qubits used in the core network system. This would reduce the complexity and time while keeping the security of classical bits intact.

Acknowledgment. The authors would like to acknowledge the time resource provided by Centre for Development of Telematics, Bengaluru, India.

References

1. Nielsen, M.A., Chuang, I.L.: Quantum Computation and Quantum Information, 2nd edn. Cambridge University Press/Massachusetts Institute of Technology, Cambridge (2000)
2. Qubit Wikipedia. https://en.wikipedia.org/wiki/Qubit. Accessed 1 Nov 2020
3. Quantiki. https://www.quantiki.org/wiki/quantum-gates. Accessed 1 Nov 2020

4. QISKIT Super dense coding. https://qiskit.org/textbook/ch-algorithms/superdense-coding.html. Accessed 1 Nov 2020
5. Science Direct. https://www.sciencedirect.com/topics/computer-science/quantum-circuit. Accessed 1 Nov 2020
6. Tan, X., Cheng, S., Li, J., Feng, Z.: Quantum key distribution protocol using quantum fourier transform. In: 2015 IEEE 29th International Conference on Advanced Information Networking and Applications Workshops, Gwangiu, pp. 96–101 (2015). https://doi.org/10.1109/WAINA.2015.8
7. Kumavor, P.D., Beal, A.C., Yelin, S., Donkor, E., Wang, B.C.: Comparison of four multi-user quantum key distribution schemes over passive optical networks. J. Lightwave Technol. **23**(1), 268–276 (2005). https://doi.org/10.1109/JLT.2004.834481
8. Brassard, G., Bussieres, F., Godbout, N., Lacroix, S.: Multiuser quantum key distribution using wavelength division multiplexing. In: Proceedings of SPIE 5260, Applications of Photonic Technology 6 (2003). https://doi.org/10.1117/12.543338
9. Xue, P., Wang, K., Wang, X.: Efficient multiuser quantum cryptography network based on entanglement. Sci. Rep. **7**, 45928 (2017). https://doi.org/10.1038/srep45928
10. Quirk - A drag-and-drop quantum circuit simulator. https://algassert.com/quirk. Accessed 1 Nov 2020

Design of CoAP Based Model for Monitoring and Controlling Physical Parameters

Vishnu Kanthan Rathina Raj[✉] and Meena Belwal

Department of Computer Science and Engineering, Amrita School of Engineering, Amrita
Vishwa Vidyapeetham, Bengaluru, India
vishnukanthan.r@microchip.com, b_meena@blr.amrita.edu

Abstract. With rapid development and increasing demand of Internet of Things
(IoT) applications, effective client server system is necessary for fast communication without data loss. IoT applications generally have sensors to provide various
physical parameters and actuators to operate as per sensor input. A specialized protocol that enables this machine-to-machine communication is Constrained Application Protocol (CoAP). The CoAP shares features with HTTP such as REST
model but personalized for constrained devices like embedded micro controllers
and networks in IoT. Industries at present use commercial small computers such
as Raspberry Pi for developing IoT applications to speed up the process. This has
certain overheads like cost and fixed set of libraries. The purpose of this paper is
to present an economical and scalable solution for IoT application development.
The working prototype model comprises a bare metal microchip pic 8-bit micro
controller, Ethernet controller, infrared sensor, browser display for sensor data,
actuator to on/off the motor and integrated TCP/IP CoAP stack libraries. This
paper demonstrates the function of the proposed model and how scalability can
be achieved by connecting multiple sensors. Infrared sensor can be replaced with
any device that generates data such as blood pressure, temperature etc.

Keywords: Internet of Things (IoT) · Constrained Application Protocol
(CoAP) · HTTP · REST model · Micro controller · Sensor · Actuator

1 Introduction

IoT applications grow exponentially with availability of sensors, actuators and cost-
effective low-power processors. It is estimated that over 17 billion devices connected
worldwide in IoT, are collecting and sharing data [1]. From home automation to business
to consumers, more and more resource constrained devices are getting connected to
network controlled and monitored from remote.

IoT implementation using single board small devices has trade-off with cost and
scalability [2]. This research is focused on developing IoT system using peripheral
interface controller (PIC) 18F87K222, a bare metal 8-bit micro controller, and achieving
complete monitoring and controlling of IoT endpoints using integrated TCP/IP CoAP
libraries that are exclusively developed for this work. It receives sensor data and sends

© ICST Institute for Computer Sciences, Social Informatics and Telecommunications Engineering 2021
Published by Springer Nature Switzerland AG 2021. All Rights Reserved
N. Kumar et al. (Eds.): UBICNET 2021, LNICST 383, pp. 105–115, 2021.
https://doi.org/10.1007/978-3-030-79276-3_9

to clients across the network. It costs below 4 USD with features such as high-speed data transfer of 10 Mbps. The proposed solution is tested with different CoAP clients to ensure interoperability.

This model would be useful in numerous real-time applications. Following is a typical application example. Getting alert notification from a remote server when CCTV camera captures image of an intruder. In this case CCTV camera is the sensor, and this model can send alert signal to a remote client upon capturing images of intruders. Instructions can also be sent from remote to control motor.

Bare metal provides convenience of direct programming of hardware registers and eliminates fixed set of libraries. Also, complete flash memory is available for programming and other important resources like CPU, memory and storage units can be utilized at their full capacity and hence it provides maximum performance and speed.

Keeping in mind low overhead, limited RAM/ROM of constrained environments, CoAP web protocol is designed with additional feature that it can communicate to HTTP as well.

The rest of the paper is organized in the following order. Section 2 presents relevant work published by various authors. Section 3 explains overview and stages of IoT and comparison between IoT and web stacks. Design and development of the proposed model is explained in Sect. 4. Section 5 presents the execution results.

2 Related Work

Louis COETZEE [3] et al. found through experiments that CoAP is an efficient transport in low signal strength environments. Dejana Ugrenovic [4] et al. analyzed and simulated the IoT healthcare system that monitors patient's health condition. Gyuhong Choi et al. [5] presented an efficient communication scheme for streaming services in IoT. Kavitha. B. C and Vallikannu. R [6] proposed Raspberry Pi based IoT for monitoring parameters and send the data to cloud. Mrs. Vaishali Puranik et al. [7] et al. discussed automation of agricultural processes such as maintenance, pest control etc. using IoT. G. Tanganelli, C. Vallati, E. Mingozzi [8] have introduced an open source CoAP library called CoAPthon, which can be used to develop IoT applications based on CoAP. This library has been used for this research work. Er. Pooja Yadav [9] et al. discussed general challenges in IoT applications in India. Likith bhushan H N and Niranjan K R [10] analysed the use of wireless sensor networks based IoT in health care applications. Dweepayan Mishra et al. [11] proposed Arudino kit based IoT programmed water system for irrigation. Madhusu-dan Singh et al. [12] described the infrastructure of IoT for Blockchain network based IoT. Dr.Digvijaysinh Rathod [13] et al. demonstrated that CoAP proxy is vulnerable and succeeded to show that the information is transiting in clear text format which is suscep-tible for attacks or manipulation of data. Ahmad Zainudin et al. [14] compared CoAP and MQTT protocols and observed that CoAP performance is better in terms of delay. Ravi Kishore Kodali et al. [15] used ESP32 and ESP8266 to implement CoAP based home automation. Jiahao Li, Gengyu Wei [16] suggested blockchain based security for CoAP nodes. Girum Ketema Teklemariam et al. [17] illustrated RESTful interface tool called RESTlets in IoT applications with sensors and actuators. Syed Roohullah Jan et al. [18] explained and compared CoAP with HTTP. Lucas R. B. Brasilino, Martin

Swany [19] presented various security threats in IoT devices. Alok Kumar Gupta et al. [20] discussed an energy saving electrical device Surveillance and Control system based on IOT. A. Paventhan et al. [21] presented an approach to using CoAP for agriculture monitoring. Matthias Kovatsch [22] proposed a system architecture for CoAP based IoT cloud services.

3 CoAP Protocol

As depicted in Fig. 1, Overview of IoT has three stages Sensors that capture data, a platform or bridge to communicate data using protocols and networks, and applications that use data.

Fig. 1. Stages of IoT

Transmitting sensor data across devices indeed is a complex task. TCP/IP stack deals with connectivity to the Internet. CoAP protocol in Application layer handles communication among devices as shown in Fig. 2.

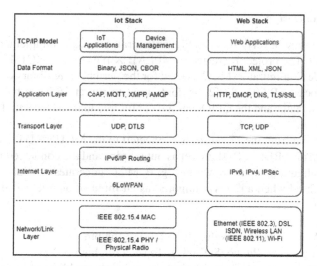

Fig. 2. Comparison of IoT & web stacks

CoAP has lightweight connectionless User Datagram Protocol (UDP) as its transport layer protocol whereas Transmission Control Protocol (TCP) is used by HTTP.

4 Methodology

4.1 High-Level Design

The high-level design of proposed model is shown in Fig. 3. Sensors and actuators are hosted as CoAP resources in the MCU. The MCU is interfaced with the ethernet controller through SPI and acts as the CoAP server. A PC/laptop acts a CoAP client that is running a client software such as CoAPthon3 or Copper. The client and the server are both connected via a conventional twisted pair ethernet cable. The client can retrieve data from the sensor and also control the state of the actuator through the Request-Response model.

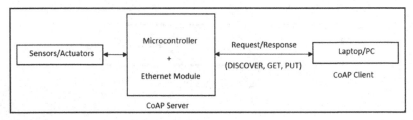

Fig. 3. High-level design

4.2 Low-Level Design

The Fig. 4 shows low-level design and Fig. 5 illustrates the actual implementation of the proposed model. The project is created in MPLAB X IDE for the PIC18F87K22 MCU. Microchip Code Configurator (MCC) is used for the peripheral & library configurations. The system clock of PIC18F87K22 is configured to run at 16 MHz by choosing the Internal RC-Oscillator and the GPIO pins are configured to interface with IR sensor and the Motor control.

The DOUT pin from the IR sensor is connected to the MCU's RD4 that is, pin.no 4 of PORTD. The pins RD0 & RD1 are set as output pins and are connected to the L293D driver. The configurations are shown in Fig. 6. MCC generates various functions that can be readily used when a pin is configured as an input or output. Some of functions used in the project are:

- Pin.no_GetValue()- Gets the value (high/low) of the pin
- Pin.no_SetHigh() and Pin.no_SetLow()- To set a pin high or low
- Pin.no_Toggle()- Toggles the current state of the pin (from high to low or low to high).

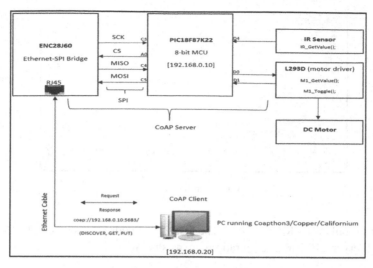

Fig. 4. Low-level design

Fig. 5. Actual implementation

4.3 Configuring TCP/IP Lite Stack

The TCP/IP lite stack version 2.2.12 (latest) has been used for implementation. The protocols UDP and ICMP are checked. The rest of them are left unchecked as it is not required for this work and can be added for future extension. But adding too many protocols like the TCP, NTP and DNS may not work as expected due to the embedded

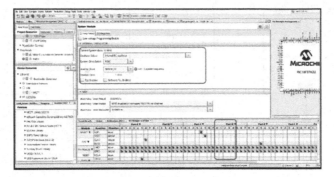

Fig. 6. Initial setup

constraints as only 8-bit MCU is used. The local IPV4 address is set as "192.168.0.10". This is the server's IP. The configurations are as shown in Fig. 7.

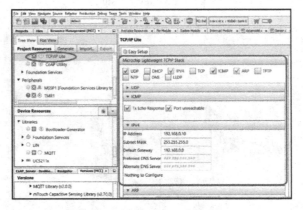

Fig. 7. TCP/IP lite configuration

4.4 Configuring CoAP Library

As the IR sensor and the DC motor are added as simple GPIO interfaces, even though the actual hardware is connected, they needed to be added as CoAP Resources. From the list of libraries, the CoAP Utility library is included to the project. The resources are added and renamed and included under a root called "Node" as shown in Fig. 8.

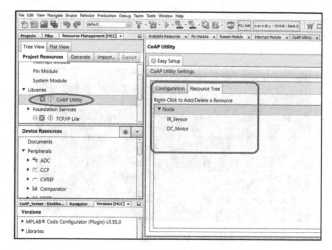

Fig. 8. CoAP utility configuration

4.5 Development of the Firmware

The main.c is as shown below Fig. 9:

```
#include "mcc_generated_files/mcc.h"
void main(void)
{
    SYSTEM_Initialize();
    INTERRUPT_GlobalInterruptEnable();
    INTERRUPT_PeripheralInterruptEnable();
    IO_RD0_SetLow();//RD0 connected to push button(open drain) which is also connected to an LED, set it low
    //To turn the motor OFF initially
    M1_SetHigh();//RD0 motor pin is set high
    M1_SetHigh();//RD1 motor pin is set high

while(1)
    {
        Network_Manage();
    }
}
```

Fig. 9. main.c

The SYSTEM_Initialize() method initializes the interrupt registers, pin manager, the system clock oscillator, timer, initiates the stack and also the CoAP resources. The Network_Manage() is called inside an infinite loop so that when the system is powered ON, the stack keeps running. It is inside this method, where all the above layers TCP/IP and the CoAP stack are managed. When GET request is performed, inside the MCU, it hits the below function SensorGetter(idx). As each resource is assigned a unique identifier, the value of the sensor data is retrieved from the configured input pin of the MCU. The working is similar for the PUT request as well, for which the command is sent to the output pins at which the actuators are interfaced.

5 Results and Discussion

5.1 PING

The ping command comes under the ICMP protocol. It is used to test if a node is able to reach out to another node in a network. The ping command is typed at the command prompt window of client laptop along with the server's IP address. As seen from the Fig. 10 the average time to hit the server and back is just 6ms.

```
C:\Windows\system32\cmd.exe
C:\Users\rvish>ping 192.168.0.10

Pinging 192.168.0.10 with 32 bytes of data:
Reply from 192.168.0.10: bytes=32 time=9ms TTL=64
Reply from 192.168.0.10: bytes=32 time=6ms TTL=64
Reply from 192.168.0.10: bytes=32 time=6ms TTL=64
Reply from 192.168.0.10: bytes=32 time=6ms TTL=64

Ping statistics for 192.168.0.10:
    Packets: Sent = 4, Received = 4, Lost = 0 (0% loss),
Approximate round trip times in milli-seconds:
    Minimum = 6ms, Maximum = 9ms, Average = 6ms
```

Fig. 10. PING command

5.2 DISCOVER

The Copper client is opened in the Chrome browser and the server's URL is entered. The default port number is 5683. After entering the URL, the server's IP is selected and the DISCOVER button on the top bar is clicked which retrieves information about the CoAP resources hosted by the server as shown in the Fig. 11.

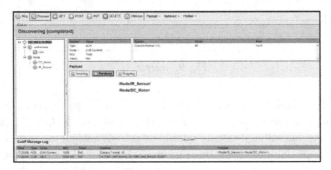

Fig. 11. DISCOVER CoAP resources

5.3 Monitoring the IR Sensor

To monitor the sensor data, the resource IR Sensor under node is selected and then the GET button is clicked. Then an obstacle is placed in front of the IR sensor and then the GET button is clicked again to verify if an obstacle detected message is received. The results are shown in Fig. 12.

Fig. 12. Monitoring Sensor data

5.4 Monitoring and Controlling the DC Motor

In order to test out the motor, the resource DC Motor is selected under Node and GET button is clicked. Initially the "Motor is OFF" message is received. When PUT button is clicked, it toggles the RD0 pin that is connected as one of the two control signals that are both set as high initially, turns on the motor. When GET button is clicked, the message "Motor is running" is received. The results are shown in Fig. 13.

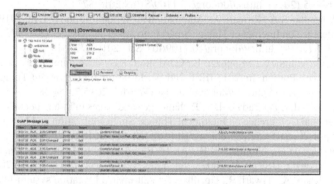

Fig. 13. Monitoring & controlling DC motor

6 Conclusion and Future Work

This paper discussed the development of an IP-address based working prototype model for Server-Client communication architecture using TCP/IP CoAP integrated libraries to monitor and control physical parameters over a local network. The client can identify the resources that are hosted in the server. This work can be further extended to cover security related issues since CoAP is susceptible to DDoS attacks. Hence, implementation of security layer such as IP-Sec or DTLS can be the next objective by taking into consideration the available flash memory and encryption-decryption overheads. Several such developed servers can also be connected to a router or a switch to form a network and deployed in an environment that needs monitoring. A client connected to such a network can easily identify a server/node with its assigned IP address.

References

1. Statista Research Department, 27 November 2016. https://www.statista.com/statistics/471 264/iot-number-of-connected-devices-worldwide/
2. Mishra, A., Reichherzer, T., Kalaimannan, E., Wilde, N., Ramirez, R.: Trade-offs involved in the choice of cloud service configurations when building secure, scalable, and efficient Internet-of-Things networks. Int. J. Distrib. Sens. Netw. **16**(2), 1–13 (2020). https://doi.org/10.1177/1550147720908199
3. Louis Coetzee: An Analysis of CoAP as Transport in an Internet of Things Environment. In: Gaberone, Coetzee, B.L., Oosthuizen, D., Mkhize, B. (eds.) IST-Africa 2018, pp. 1–7 (2018)
4. Ugrenovic, D., Gardasevic, G.: CoAP protocol for web-based monitoring in IoT healthcare applications. In: 2015. 23rd Telecommunications Forum (TELFOR), Belgrade, pp. 79–82 (2015)
5. Choi, G., Kim, D., Yeom, I.: Efficient streaming over CoAP. In: 2016 International Conference on Information Networking (ICOIN), Kota Kinabalu, pp. 476–478 (2016). https://doi.org/10.1109/ICOIN.2016.7427163
6. Kavitha, B.C., Vallikannu, R.: IoT based intelligent industry monitoring system. In: 2019 6th International Conference on Signal Processing and Integrated Networks (SPIN), Noida, India, pp. 63–65 (2019). https://doi.org/10.1109/SPIN.2019.8711597
7. Puranik, V., Ranjan, S.A., Kumari, A.: Automation in Agriculture and IoT. In: 2019 4th International Conference on Internet of Things: Smart Innovation and Usages (IoT-SIU), Ghaziabad, India, pp. 1–6 (2019). https://doi.org/10.1109/IoT-SIU.2019.8777619.
8. Tanganelli, G., Vallati, C., Mingozzi, E.: CoAPthon: Easy development of CoAP-based IoT applications with Python. In: 2015 IEEE 2nd World Forum on Internet of Things (WF-IoT), Milan, pp. 63–68 (2015).https://doi.org/10.1109/WF-IoT.2015.7389028
9. Yadav, E.P., Mittal, E.A., Yadav, H.: IoT: challenges and issues in Indian perspective. In: 2018 3rd International Conference on Internet of Things: Smart Innovation and Usages (IoT-SIU), Bhimtal, pp. 1–5 (2018). https://doi.org/10.1109/IoT-SIU.2018.8519869
10. Likith Bhushan, H.N., Niranjan, K.R.: An IP Based Patient Monitoring Smart System in Hospitals. Paper ID: IJSRDV4I120073, vol. 4, no. 12, pp. 154–156, 1 March 2017
11. Mishra, D., Khan, A., Tiwari, R., Upadhay, S.: Automated irrigation system-IoT based approach. In: 3rd International Conference on Internet of Things: Smart Innovation and Usages (IoT-SIU), Bhimtal, pp. 1–4 (2018). https://doi.org/10.1109/IoT-SIU.2018.8519886
12. Singh, M., Singh, A., Kim, S.: Blockchain: agame changer for securing IoT data. In: 2018 IEEE 4th World Forum on Internet of Things (WF-IoT), Singapore, pp. 51–55 (2018).https://doi.org/10.1109/WF-IoT.2018.8355182
13. Digvijaysinh, R.: Security analysis of constrained application protocol (CoAP): IoT protocol **6**(8), 37 (2017)
14. Zainudin, A., Syaifudin, M.F., Syahroni, N.: Design and implementation of node gateway with MQTT and CoAP protocol for IoT applications. In: ICITISEE, Yogyakarta, Indonesia, pp. 155–159 (2019). https://doi.org/10.1109/ICITISEE48480.2019.9003734
15. Kodali, R.K., Yatish Krishna Yogi, B., Sharan Sai, G.N., Honey Domma, J.: Implementation of home automation using CoAP. In: TENCON 2018 - 2018 IEEE Region 10 Conference, Jeju, Korea (South), pp. 1214–1218 (2018). https://doi.org/10.1109/TENCON.2018.8650135

16. Li, J., Wei, G.: Research on CoAP resource directory based on blockchain. In: 2020 IEEE 4th Information Technology, Networking, Electronic and Automation Control Conference (ITNEC), Chongqing, China, pp. 1730–1735 (2020). https://doi.org/10.1109/ITNEC48623.2020.9084996

17. Teklemariam, G.K., Hoebeke, J., Van den Abeele, F., Moerman, I., Demeester, P.: Simple RESTful sensor application development model using CoAP. In: 39th Annual IEEE Conference on Local Computer Networks Workshops, Edmonton, AB, 2014, pp. 552–556 (2014). https://doi.org/10.1109/LCNW.2014.6927702

18. Roohullah, S., Fazlullah, K., Farman, U, Nazia, A, Muhammad, T.: USING CoAp protocol for resource observation in IOT. Int. J. Emerg. Technol. Comput. Sci. Electron. (IJETCSE) **21**(2), 385–388 (2016). ISSN: 0976-1353

19. Brasilino, L.R.B., Swany, M.: Mitigating DDoS Flooding attacks against IoT using custom hardware modules. In: 2019 Sixth International Conference on Internet of Things: Systems, Management and Security (IOTSMS), Granada, Spain, pp. 58–64 (2019). https://doi.org/10.1109/IOTSMS48152.2019.8939176

20. Gupta, A.K., Johari, R.: IOT based electrical device surveillance and control system. In: 2019 4th International Conference on Internet of Things: Smart Innovation and Usages (IoT-SIU), Ghaziabad, India, pp. 1–5 (2019). https://doi.org/10.1109/IoT-SIU.2019.8777342

21. Lavanya, P., Sudha, R.: A study on WSN Based IoT application in agriculture. In: 2018 3rd International Conference on Communication and Electronics Systems (ICCES), Coimbatore, India, pp. 1046–1054 (2018). https://doi.org/10.1109/CESYS.2018.8724020

22. Kovatsch, M., Lanter, M., Shelby, Z.: Californium: scalable cloud services for the Internet of Things with CoAP. In: 2014 International Conference on the Internet of Things (IOT), Cambridge, MA, pp. 1–6 (2014). https://doi.org/10.1109/IOT.2014.7030106

Hardware Trojan Detection Using XGBoost Algorithm for IoT with Data Augmentation Using CTGAN and SMOTE

C. G. Prahalad Srinivas[✉], S. Balachander, Yogesh Chandra Singh Samant,
B. Varshin Hariharan, and M. Nirmala Devi

Department of Electronics and Communication Engineering, Amrita School of Engineering,
Coimbatore, Amrita Vishwa Vidyapeetham, Coimbatore, India
cb.en.u4ece17239@cb.students.amrita.edu, m_nirmala@cb.amrita.edu

Abstract. Internet of things is the next blooming field in engineering. Seamless connectivity has become the new normal when devices are manufactured. This demand for IoT (Internet of Things) also results in various threats. Hardware trojans are one such threat which is encountered. Hardware trojan, being sneak circuits are overlooked when a new chip is manufactured. This further solidifies the threat posed by the hardware trojans. Applying machine learning model directly to the dataset will prove futile, since the amount of trojan infected nets are very minimal compared to the regular nets. In this work, this problem is addressed by creating a machine learning model, which can handle the data imbalance, using CTGAN (Conditional Tabular Generative Adversarial networks) and SMOTE (Synthetic Minority Oversampling Technique) algorithms, and detect the trojan infected nets.

Keywords: Hardware Trojan · SMOTE · Machine learning · CTGAN · XGBoost

1 Introduction

Hardware trojans are sneak circuits which can be of different types and can perform different functions. These trojans are also occupy minimal area and hence are incredibly hard to detect manually. The insertion of these trojan can happen at several stages during fabrication. This increases the difficulty in detecting the presence of a trojan in a chip.

In this work, an approach using machine learning is suggested to solve this problem. Despite having neural network and deep learning techniques, machine learning algorithms were chosen for the detection because of its speed of training and execution. Since this work deals with IoT (Internet of Things), the detection needs to happen almost instantaneously, and training also needs to be done very quickly if a new circuit is being introduced. This is the main reason for choosing machine learning algorithms over neural networks.

The model that has been chosen for hardware trojan detection is XGBoost (eXtreme Gradient Boosting). It is currently considered one of the best machine learning models.

N. Kumar et al. (Eds.): UBICNET 2021, LNICST 383, pp. 116–127, 2021.
https://doi.org/10.1007/978-3-030-79276-3_10

It has also been proven to be very effective for the problem at hand [1, 2]. XGBoost was designed for its swiftness over the normal gradient boosting algorithm and is able to consistently outperform other models.

Another big problem faced when it comes to trojan detection is the lack of significant trojan nets. Since the number of nets are too small, the model gives more importance to the non-trojan nets and hence is not able to detect the trojan nets accurately. This is why two of the widely used synthetic data generation techniques are being tested and evaluated, namely CTGAN (Conditional Tabular Generative Adversarial networks) and SMOTE (Synthetic Minority Oversampling Technique).

Finally, the results of CTGAN and SMOTE are being compared. For the purpose of this comparison, the main metric is taken to be recall and not accuracy. The reason for this being that recall is able to convey how many relevant data points are available in the dataset. Since the data is being synthetically generated, recall will provide a good measure for comparison.

This work starts by having a look at previous solutions for the problem. A basic understanding of hardware trojans and the types of hardware trojans used for this work is then provided. Next, the methodology of the experiments conducted is discussed. It includes the feature extraction process, the data augmentation process and the machine learning model building process. Finally, the results obtained by the experiment is analyzed and the final conclusion is drawn.

2   Related Works

Similar Approaches for Different Problems
In the [1] by Richa Sharma et al., they have suggested a Hardware trojan detection technique using Weighted XGBoost Classifier. The authors have extracted observability and controllability measures for all the nets and have suggested that the trojan nets display a higher observability and controllability measure. They have also suggested a permutation-based feature selection for this problem, which was found to be interesting. Cheng Dong et al. [2] have taken an approach of hardware trojan detection towards the IoT field. They have suggested that the IoT is the next big thing and hardware chips and ICs will be a very integral part of this process. They have used XGBoost for hardware trojan detection and have achieved an accuracy of 99.83%. The paper [3] provides an overview on the XGBoost algorithm and the mathematics behind XGBoost.

In this work, a different set of features has been considered to try and improve the accuracy. Synthetic data generation will also be used to try and improve the dataset and hence improve the accuracy of the model. This work was tested with 7 different circuits with 3 different trojans inserted to ensure that this approach produces consistent results for a wide range of circuits.

Different Approaches for Similar Problems
In [4], the author detects trojans using a compressive sensing algorithm to detect hardware trojans. The author talks about generating test patterns which can effectively trigger the trojans into discharging their payload. This method can be considered as a very good

replacement for the golden chip method of trojan detection. In [5] the author is suggesting a way to detect trojans using weighted average voting. Here, the authors send the same set of inputs to different modules of the ISCAS (IEEE International Symposium on Circuits and Systems) circuits and then considers the history of these outputs to detect if a trojan is present or not. [6] Discusses about detecting trojans using bitstreams of output data obtained by using different modules of the same circuit. If there were any viruses, then the bitstream is going to become abnormal, hence showing that the particular circuit has an infection. [7] uses leakage power calculation for multiple iterations of the circuit to observe potential changes and subsequently detect the hardware trojan inserted circuits. Even though the above methods can successfully detect trojans, they will not be able to point to the exact net that the trojan is present.

3   Hardware Trojans and Its Types

Hardware trojans are small bits of circuit that can decrease the performance of a given circuit. These trojans are sometimes also used to steal valuable information or data.

3.1   Threat Model

Trojans need two specific ingredients for it to function properly. One of them is a trigger and another is a payload. A trigger in a trojan is a circuit which will turn on the trojan at the right point of time. It is important for hardware trojans to not be detected during testing phase. That is why the trigger circuit will wait for the right condition to trigger the trojan circuit. This right condition will depend on the designer who is designing the trojan. Some triggers will wait for a right set of inputs, while some may trigger after certain amount of time.

The payload of a trojan is the part of the trojan circuit which actually causes harm to the host circuit. These can vary from changing the functionality of the circuit to extending the time taken to produce the desired output. The trojans can be vastly classified based on the types of triggers and the types of payload [8]. Gate level netlist is considered as an input, and 3 types of trojans are analyzed.

3.2   Trojans in IoT

The significance of hardware trojans on IoT is that most of IoT systems are realized using hardware deployment. And where there is hardware, there is always the risk of having a hardware trojan. Starting from a small LoRa (Long Range) module to a node MCU (Micro Controller Unit), a hardware trojan can be present everywhere. Since these products are bought by the customers and the distributors have a hard time checking and removing the trojans, the trojans inserted by a manufacturer may go unnoticed. That is why, in this work, Trojan detection will be made possible with a cheap yet effective machine learning based method.

3.3 The Types of Trojans Considered for Testing

Three types of hardware trojans are considered for detection. Namely: Always on, sequential and combinational trojan.

Always on Trojans
An always on trojan is, as the name suggests, always on. These are the only type of trojans that don't require a trigger circuit to turn them on. Thus, identifying the trojans is a bit more difficult because the amount of circuitry involved can be as small as a single gate.

Sequential Trojans
Sequential trojans [9] are the trojans which are activated using a sequential circuit. This circuit is commonly an asynchronous up counter which will trigger the circuit when it reaches a certain value, or after a certain number of a single operation is performed. Another example of a sequential trojan is something known as a "time bomb". These trojans keep track of time and once the time is reached, they release the payload into the host.

Combinational Trojans
Combinational trojans are trojans which are activated using a combinational circuit. Common examples of combinational trojans are FSM (Finite State Machine) trojans. These trojans have an FSM built as a trigger and will trigger the payload only if a certain pattern of inputs is received. The probability of receiving that specific pattern during the testing phase is very small and hence these trojans are effective at avoiding detection during testing phase.

4 Methodology

4.1 Feature Extraction

Several features have been extracted from the gate level netlists. Seven different circuits with 3 different trojans are considered. Table 1 contains the details of the benchmark circuits which were considered.

The features which have been extracted from these circuits are described in Table 2.

The dataset with which this experiment is done, in general has two types of features the first one is discrete and the other is continuous. In the datasets which are considered, the following has been observed:

Features like Level, Connectivity, Po, Score, Fan-in, Load, Resistance, Pins, Total net are considered as discrete features, since the level of the net from the input layer in any given circuit comes from fixed set of values. For connectivity, the number of gates the given net is connected, comes from fixed set. For Po, the number of levels the net is away from the output net, comes from the fixe set similar to Level. Score is summation of connectivity and Po, hence it is also considered as discrete feature. The same applies to Fan-in, and Fan-out. Generally, a net consists of combination of gates, hence each gate has its resistance and load resistance. Hence Resistance and Load resistance are

Table 1. List of benchmark circuits and the type of inserted trojan

Circuit name	Always on trojan	Sequential trojan	Combinational trojan
C17	Yes	Yes	Yes
C432	Yes	Yes	Yes
C1908	No	Yes	Yes
C6288	Yes	Yes	Yes
S27	Yes	Yes	Yes
S298	Yes	Yes	Yes
S820	Yes	Yes	Yes

Table 2. List of features extracted

Feature Name	Description
Level	The level of the net from the input layer
Connectivity	The number of gates the given net is connected to
Po	The number of levels the net is away from the output nets
Score	The sum of level, connectivity and po
Fan-in	The number of incoming nets into the gate
Load	Capacitive load of the net
Resistance	Resistive load
Pins	Number of pins
Total net	Maximum nets available
Static Prob	Probability that a certain net will stay in the current state
Toggle rate	Probability that a certain net will switch from current state
Switching power	The power needed to switch the net from one state to another

considered as discrete features. And Total resistance is sum of Load resistance and circuit resistance, again it is also considered as discrete feature.

But features like Static probability, Toggle rate and switching power are considered as continuous. Since static probability means the time duration for which the net will be at logic '1'. It varies from net to net and hence it doesn't come from a set of values. Toggle rates means the number of times the signal will change its state, hence it doesn't take any fixed set of values. Switching power too doesn't come from a fixed set of values. Hence the discussed 3 features can be collectively grouped as Continuous features.

4.2 Handling Data Imbalance Problem

A distinct problem was encountered when extracting the features. That problem is that the amount of nets which do not have trojan far outweigh the nets which are trojan infected. This creates a data imbalance problem, that is, the amount of data for one particular class is greater than that of the other class.

To mitigate this problem, two algorithms are being used: SMOTE Algorithm and CTGAN algorithm. These two algorithms are used to augment the dataset so that the amount of trojan infected and trojan free nets become the same.

CTGAN
Conditional Tabular Generative Adversarial Network is method of modelling tabular data by involving mode specific normalization, which addresses Non- Gaussian and multimodal distribution of continuous columns and conditional generator and training by sample, which address the imbalanced discrete columns [10].

Mode Specific Normalization
Each continuous column is modelled using Variational Gaussian Mixture Model to estimate the number of modes and fit a mixture model. For each value in a continuous column, the probability of that value coming from each mode is calculated. Sample one mode from the given probability density and used that mode to normalize the value. Then a row could be expressed as a concatenation of continuous and discrete columns.

$$N_{i,j} = \frac{(c_{i,j} - \eta_n)}{4\,\Phi_n} \tag{1}$$

Where,

$N_{i,j}$ is the Normalized value at i^{th} column and j^{th} row.
$c_{i,j}$ is the value at i^{th} column and j^{th} row.
η_n is the nth Gaussian mode in i^{th} column.
Φ_n is the standard deviation of nth Gaussian distribution in i^{th} column.

Conditional Generator and Training by Sample
In a traditional GAN (Generative Adversarial Network) approach, the vectored input is fed to the discriminator without taking into consideration for the imbalance in the discrete columns (categorical columns). While sampling the data when rows of one particular category isn't sufficiently represented, the generator won't give the desired output, i.e. the generator is not trained correctly. The generator can be interpreted as the conditional distribution of rows given that particular value at that particular column,

$$s \sim P(row|D_i = v), \tag{2}$$

The original data distribution can be reconstructed as follows

$$P(row) = \sum_{val \in D_i} P(row|D_i = v)P(D_i = val) \tag{3}$$

Where,

s is the generated sample,

D_i is the ith discrete column and.

v is value at the ith discrete column.

The output produced by the generator is to be assessed by the discriminator. The sampling of real training data and construction of the condition vector (concatenation of one hot encoded of discrete columns based on the requirements) should comply so that the discriminator could estimate the distance. A good sampled condition vector and training data could almost explore all the values in discrete columns.

SMOTE

Synthetic Minority Oversampling Technique (SMOTE) is a technique used to address the data imbalance problem by oversampling the minority class. This data augmentation method works on the principle of nearest neighbours, where new data points will be created by combining two or more original instances. When a random datapoint a is being considered from the minority class on the feature space, the k of the nearest neighbours is found for that particular datapoint. A random nearest neighbour b is selected, and an imaginary line l is drawn in-between the two points a and b. New instances are added or sampled on that line l, as these new synthetic datapoints are a combination of a and b. This process is repeated for k different datapoints that are near to a.

By expanding this process to the other instances in the minority class, multiple duplicates can be created, thus solving the data imbalance setback. This algorithm is effective because of the fact that the newly created instances will be closer the existing instances in the feature space. This helps the classifier to build a larger decision region that contains the datapoints belonging to the minority class.

For the dataset that is being considered for this application, 15 different features are present in it. The above concept will be expanded to a 15-dimensional vector space, where a specific a and b will correspond to a single datapoint, and will be placed in a unique position. However, the type of each feature can be either continuous or discrete as mentioned in the feature extraction section. When the oversampling takes place, the continuous feature values are taken as such, since they carry the freedom to have a value chosen in-between two samples. This isn't the case with the discrete values, since it is rigid as the feature carries specific values. To overcome this, SMOTE considers the discrete values as continuous while oversampling, but rounds off the oversampled values to the nearest integer when it comes to discrete features.

The realization of this method is achieved in python with the help of *imblearn* library, which contains the model for SMOTE under the sub-library *over_sampling*. The dataset with all the features that are both continuous and discrete was pushed into the function as a whole. The function has the capability to identify the minority class and the number of samples required to balance the dataset. Also, the neighbour value k can be tuned accordingly on top of the default value to produce the required number of samples. This advantage was used to create appropriate samples in the dataset containing limited samples, as the value k was minimized.

Given below are the results obtained after applying SMOTE algorithm to the considered dataset (Table 3).

Table 3. Comparison of instances before and after applying SMOTE

Circuit	Trojan type	Instances before SMOTE			Instances after SMOTE		
		0	1	Total	0	1	Total
C17	Always on	7	9	16	9	9	18
	Combinational	5	8	13	8	8	16
	Sequential	6	11	17	11	11	22
C432	Always on	116	85	201	116	116	232
	Combinational	86	113	199	113	113	226
	Sequential	107	88	195	107	107	214
C1908	Combinational	632	283	915	632	632	1264
	Sequential	706	208	914	706	706	1412
C6288	Always on	2133	320	2453	2133	2133	4266
	Combinational	2035	413	2448	2035	2035	4070
	Sequential	2251	198	2449	2251	2251	2502
S27	Always on	11	10	21	11	11	22
	Combinational	7	11	18	11	11	22
	Sequential	10	12	22	12	12	24
S298	Always on	92	49	141	92	92	184
	Combinational	86	53	139	86	86	172
	Sequential	100	41	141	100	100	200
S820	Always on	253	55	308	253	253	506
	Combinational	246	63	309	NA*		
	Sequential	264	49	313	264	264	528

*Note that S280 did not produce synthetic data due to the unavailability of nearest neighbours.

XGBoost Algorithm

eXtreme Gradient Boosting algorithm has been used for training the model over the dataset. This algorithm comes under the umbrella of ensemble models. This machine-learning model are algorithm which uses a weaker classifier to make itself stronger. There are broadly two types of ensemble algorithms, bagging and boosting. The XGBoost algorithm comes under the boosting category. It is a decision tree-based ensemble machine learning technique that uses the Gradient boosting framework for machine learning prediction.

XGBoost tries to find the optimal output value for a tree f_t in an iteration t that is added to minimize the loss function across all data point (4) [3].

$$L(t) = \sum_{i=1}^{n} l\left(y_i, \ \hat{y}^{(t-1)} + f_t(x_i)\right) + \Omega(f_t) \tag{4}$$

Where,

l represents a differentiable convex loss function
$\hat{y}_i$ is the prediction
y_i is the target

The steps to perform XGBoost are as follows:

1. For Initial predictions and the corresponding initial residual errors, the mean of target values is calculated.

$$h_0(x) = \text{Mean}(Y) \tag{5}$$

$$\hat{Y} = Y - h_0(x) \tag{6}$$

Where,

h_0 represents the initial prediction
$\hat{Y}$ represents the initial residual errors
Y represents the target values

2. A model is trained with independent variables and residual errors which is used to predict the output.
3. The output obtained from the model is used for additive predictions and the calculation of residual errors with some learning rate as shown in the following images.

$$\text{Gain} = \text{Left}_{\text{similarity}} + \text{Right}_{\text{similiarity}} - \text{Root}_{\text{similarity}} \tag{7}$$

$$Similarity\ Score = \frac{\left(\sum Residual_i\right)^2}{\sum \left[previous\ probablity_i \times (1 - previous\ probablity_i)\right] + \lambda} \tag{8}$$

$$Output = \frac{\left(\sum Residual_i\right)}{\sum \left[previous\ probablity_i \times (1 - previous\ probablity_i)\right] + \lambda} \tag{9}$$

Where,

λ represents the regularization factor.

4. Steps 2 and 3 are repeated for M times until the specified number of models are built.
5. The final prediction from boosting is that adding the sum of all previous predictions made by the models.

$$F_{final} = h_0(x) + \alpha \sum h_i(x) \tag{10}$$

Where,

F_{final} is the final prediction

$h_0(x)$ is the initial prediction

$h_i(x)$ is the previous predictions

α is the learning rate

XGBoost is a very reliable, powerful, and sophisticated algorithm that offers the state-of-the-art results for several complex problems.

5 Results

Table 4. Results after applying CTGAN and SMOTE algorithms on the dataset

Circuit	Trojan type	Recall before augmentation (%)		Recall after SMOTE (%)		Recall after GAN (%)		Accuracy of XGBoost classifier (%)		
		0	1	0	1	0	1	Before data synthesis	After SMOTE	After CTGAN
C17	Always on	67	100	80	100	86	100	83	83	88
	Combinational	0	75	67	67	94	100	60	67	94
	Sequential	0	80	100	100	100	75	67	100	97.5
C432	Always on	94	79	94	89	93	98	87	91	97.7
	Combinational	68	58	65	71	99	66	62	68	95
	Sequential	98	96	90	97	98	100	97	93	100
C1908	Combinational	99	95	98	98	100	96	97	98	100
	Sequential	98	97	98	97	98	100	98	98	100
C6288	Always on	97	57	93	94	99	99	92	93	99
	Combinational	95	58	94	86	88	100	90	90	98
	Sequential	100	17	88	93	100	100	94	91	100
S27	Always on	50	100	60	100	100	100	71	75	100
	Combinational	100	75	80	100	97	75	83	88	95
	Sequential	75	100	100	40	100	50	88	62	98
S298	Always on	94	67	83	77	89	99	85	80	98
	Combinational	75	83	75	67	100	97	78	70	97
	Sequential	88	60	94	88	92	99	79	91	99
S820	Always on	91	60	91	94	99	100	81	92	100
	Combinational	95	28	NA		97	99	83	NA	98
	Sequential	96	54	87	82	96	100	90	85	99

From the above results, it can be concluded that the CTGAN is better than the SMOTE algorithm for the problem on hand (Table 4). It can also be noticed that the SMOTE algorithm was unable to generate synthetic data for S820 combinational trojan circuit. This was because there was not sufficient minority neighbours for SMOTE algorithm to work on. The CTGAN algorithm on the other hand is robust and is able to overcome the shortcoming of the SMOTE algorithm.

This is one of the first works which use recall as a metric for augmentation. Recall explains about how many relevant items are being selected in the dataset. True positive explains when a sample from class A is predicted to be in class A, and when the same sample is predicted to be in class B it is false negative. Obtaining a recall of 100% reveals that there are no false negatives, and all the samples that are being selected are relevant to its particular class.

Since, after all, augmentation technique introduce new datapoints, it is vital for the datapoints to be relevant to the particular class. This is the reason why recall was chosen for evaluation. The higher the recall value, the more relevant the generated datapoints are.

The average recall has been calculated and tabulated in Table 5.

Table 5. Comparison of average recall values

	Before augmentation	SMOTE	CTGAN
Average Recall 0	79	86.157	96.25
Average Recall 1	71.95	86.315	92.65

From Table 5, it can be clearly observed that the average recall of CTGAN is far better than the average recall of SMOTE. Although SMOTE has made a huge jump in average recall when comparing with the recall before augmentation, CTGAN has achieved an even higher recall than SMOTE.

Similarly, the average accuracy before augmentation was 83.25% and after using SMOTE was 85% and with CTGAN, it was at its highest at 97.66%, with 6 circuits achieving a perfect 100% accuracy.

6 Conclusion

From the above results, it can be concluded that resolving the data imbalance problem is improving the accuracy and the recall of the model. These trojan detection techniques can be applied to any chips and when CTGAN approach is used, accuracy of almost 100% can be achieved. Also once trained and deployed, this model can be used over and over for testing IoT chips. This helps the distributors to detect trojans with very less resources, time and money and help the customers by providing them trojan free ICs (Integrated Circuits) for use in IoT.

In future this approach can be experimented for trust benchmark circuits and check if they are also able to produce good results. There is also a scope for fusing two or more models to get 100% accuracy for all the circuits.

References

1. Richa, S., et al.: A new hardware trojan detection technique using class weighted XGBoost classifier. In: 2020 24th International Symposium on VLSI Design and Test (VDAT). IEEE (2020)
2. Chen, D., et al.: A machine-learning-based hardware-Trojan detection approach for chips in the Internet of Things. Int. J. Distrib. Sensor Netw. **15**(12), 1550147719888098 (2019)

3. Tianqi, C., Guestrin, C.: Xgboost: a scalable tree boosting system. In: Proceedings of the 22nd ACM Sigkdd International Conference on Knowledge Discovery and Data Mining (2016)
4. Priyatharishini, M., Nirmala Devi, M.: A compressive sensing algorithm for hardware trojan detection. Int. J. Electr. Comput. Eng. **9**(5), 4035 (2019)
5. Aishwarya, G., et al.: Virtual instrumentation-based malicious circuit detection using weighted average voting. In: Microelectronics, Electromagnetics and Telecommunications, pp. 423–431. Springer, Singapore (2018). https://doi.org/10.1007/978-981-10-7329-8_43
6. Mohankumar, N., Jayakumar, M., Nirmala Devi, M.: CRC-based hardware trojan detection for improved hardware security. In: Microelectronics, Electromagnetics and Telecommunications, pp. 381–389. Springer, Singapore (2018). https://doi.org/10.1007/978-981-10-7329-8_39
7. Priya, S.R., et al.: Hardware malicious circuit identification using self-referencing approach. In: 2017 International Conference on Microelectronic Devices, Circuits and Systems (ICMDCS). IEEE (2017)
8. Ramesh, K., et al.: Trustworthy hardware: identifying and classifying hardware trojans. Computer **43**(10), 39–46 (2010)
9. Xinmu, W., et al.: Sequential hardware trojan: side-channel aware design and placement. In: 2011 IEEE 29th International Conference on Computer Design (ICCD). IEEE (2011)
10. Lei, X., et al.: Modeling tabular data using conditional gan. *arXiv preprint* arXiv:1907.00503 (2019)

Data Analytics and Cloud Computing

Predictive Modeling of the Spread of COVID-19: The Case of India

Sriram Sankaran[1], Vamshi Sunku Mohan[1(✉)], Mukund Seshadrinath[2],
Krushna Chandra Gouda[3], Himesh Shivappa[3], and Krishnashree Achuthan[1]

[1] Center for Cybersecurity Systems and Networks, Amrita Vishwa Vidyapeetham,
Amritapuri, India
{srirams,vamshis}@am.amrita.edu, krishna@amrita.edu
[2] Department of Computer Science and Engineering, Amrita Vishwa Vidyapeetham,
Amritapuri, India
mukund777@am.students.amrita.edu
[3] CSIR Fourth Paradigm Institute, Bangalore, India
{kcgouda,himesh}@csir4pi.in

Abstract. COVID-19 has been the most notorious pandemic affecting the entire world resulting in numerous deaths thus crippling the world economy. While vaccines are in the process of being developed for protection, countries are implementing measures such as social distancing to prevent the spread of the virus. Also, there exists a need for developing mathematical models to predict the rate of spread of COVID-19 and quantify its impact on countries such as India. Towards this goal, we developed a realistic COVID-19 dataset consisting of state-wide distribution of number of cases in India from March-July 2020. Further, we conduct exploratory data analysis on the dataset to understand the states and their corresponding growth rates. This enables us to cluster states with exponential and non-exponential growth rates as well as assess the effectiveness of lockdown imposed to curb the spread of virus. Finally, we develop predictive models using Auto-Regressive Integrated Moving Average (ARIMA) and Long Short-Term Memory Networks (LSTM) on time-series data for top-10 affected states in India to predict the rate of spread and validate their accuracy. Finally, our models can be used to guide the development of mechanisms for optimal resource allocation of healthcare systems and response.

Keywords: COVID-19 · ARIMA · LSTM

1 Introduction

Promoting the health and well-being of citizens is of paramount importance for ensuring the safety and security of the country. The recent outbreak of the COVID-19 virus has caused a significant impact across the globe by affecting the physical and mental well-being of the people. Symptoms of the virus range

N. Kumar et al. (Eds.): UBICNET 2021, LNICST 383, pp. 131–149, 2021.
https://doi.org/10.1007/978-3-030-79276-3_11

from simple cold and fever to a much more complex respiratory illness [1–4] thus requiring the use of ventilators for artificial respiration. While vaccines are being developed to make people of diverse health backgrounds resilient to the virus, a much more promising second line of defense is to promote the concept of social distancing and night curfew among people to limit the spread of the virus. Such measures can be taken only if the government has a estimate of the rise in the number of cases in the near future based on the recorded number of cases in the past.

The impact of COVID-19 varies for different countries thus requiring the use of techniques depending on the needs and requirements. Statistics show that there are about 26 million people affected by the virus due to the increase in the contact rates and about 863,000 deaths worldwide [5]. The United States occupies the topmost position in the world in terms of the number of infected people [6] followed by Brazil. While India remained at the bottom of the list initially, it soon climbed up due to the increase in the mobility of the citizens and the contact rates. Several countries impose a variety of techniques such as a mandatory 2-week institutional or home quarantine so that the symptoms of the virus wither away by the end of the quarantine.

While the impact of COVID-19 varies for different countries, there exists a need for exploratory data analysis to understand the characteristics of individual states as well as obtain a summary of the datasets. These characteristics are typically grouped for comparative analysis and represented in a pictorial manner through the use of scatterplots, bar plots, and histograms. Further, Exploratory data analysis often involves conducting statistical analysis to formulate a hypothesis that leads to conducting further experiments to validate the behavior. Thus, there exists a need for quantifying the impact of COVID-19 for countries such as India through the use of Exploratory data analysis.

While Exploratory data analysis shows promise, there exists a need for predictive modeling of a number of COVID-19 cases for each of the states in India to predict the rate of spread of the virus. Predictive modeling relies on exploratory data analysis to understand the data, analyze the trends, and obtain meaningful summarizations so that appropriate approaches can be applied. Approaches for predictive modeling range from simple linear regression to advanced techniques such as time series analysis and deep learning. Typically, predictive modeling classifies the data into training and testing sets and fits a model on the training set and evaluates the model on the testing set to determine the accuracy.

In this paper, we develop a predictive approach to modeling the impact of COVID-19 on India. First, we conduct exploratory data analysis to quantify the impact of COVID-19 on individual states in a comparative manner. In particular, we pictorially represent the top 10 states in India that are most affected by COVID-19 and analyze the growth rates. Further, we determine the effectiveness of the lockdown imposed at different time intervals for each of the states by observing the growth rates. Finally, we leverage techniques such as time series analysis and deep learning to predict the rate of spread of the virus using the number of cases for the individual states in India. These models can collectively be used to guide the development of mechanisms for optimal resource allocation and response.

This paper is structured as follows. Section 2 investigates existing work in this area. Section 3 describes exploratory data analysis with particular emphasis on the number of cases, growth rates, and the impact of lockdown on growth rates. Section 4 discusses techniques based on time series analysis and deep learning to build predictive models for COVID-19. Section 5 concludes the paper followed by references.

2 Related Work

In this section, we review existing literature for COVID-19 that are very relevant to this paper. In particular, we divide our literature review into the pertinent areas of data analysis, mathematical modeling and predictive modeling.

2.1 Data Analysis

Statistics pertaining to COVID-19 for India have been curated on a constant basis in [5] in a crowd-sourced manner. In particular, [5] reports a number of cases on a daily as well as a cumulative basis along with the analysis of trends and patterns for each of the states in India. One of the extensively used apps for tracing COVID-19 hotspots is AarogyaSetu [7] launched by the Government of India. AarogyaSetu [7] is a location-aware app that relies on Bluetooth to identify hotspots in COVID-19 infected regions. While AarogyaSetu is primarily used in India, countries have mandated the usage of different apps for identifying COVID-19 hotspots through contact tracing. While AarogyaSetu shows promise, it has significant concerns in terms of privacy [8]. Ahmed et al. [8] has presented a survey of the list of contact tracing apps available in the market along with corresponding security and privacy concerns.

2.2 Mathematical Modeling

While data analysis is of significant importance, there exists a need for developing mathematical models and projections [9] to study the impact of virus spread. Few works [10–13] have developed models for analyzing the spread of COVID-19 in different countries. Chen et al. [10] have developed epidemiological models using Susceptible-Infected-Recovered (SIR) to study the impact of virus spread in China and analyzed the impact of lockdown under different conditions. Giamberardino et al. [11] formulated the problem of optimal resource allocation of healthcare resources during a pandemic as integer linear programming and developed heuristics. Further, the authors utilize the SIR model as the input for the optimal resource allocation problem. Li et al. [12] claim that popular social networking and streaming websites have been publishing fake information pertaining to COVID-19. Jin et al. [13] modeled the impact of news and rumors on Twitter using epidemiological models. However, their study does not consider the impact of fake information propagated during COVID-19.

2.3   Predictive Modeling

Few works [14–17] have developed techniques using machine learning to combat the spread of the virus. Rustam et al. [14] developed a suite of supervised machine learning techniques for forecasting the impact of COVID-19. Jamshidi et al. [15] leveraged deep learning techniques for the diagnosis and treatment of COVID-19. Hussain et al. [16] developed AI based techniques for COVID-19. Shoeibi et al. [17] presented a review of deep learning techniques for automated detection and forecasting of COVID-19.

In contrast to existing approaches, we conduct exploratory data analysis for studying the impact of COVID-19 for India. Further, we develop predictive models using Time Series Analysis and Deep learning to predict the rate of spread of COVID-19.

3   Exploratory Data Analysis

In this section, we conduct exploratory data analysis of COVID-19 dataset containing the number of states in India. We start with describing the dataset along with analyzing the number of cases and corresponding growth rates for the individual states in India. Finally, we examine the effectiveness of the lockdown on the growth states for the individual states.

3.1   Data Description

In this work the number of positive cases of COVID-19 reported is used for the period March-Aug 2020. The data is being collected from sources like Ministry of Health and Family Welfare (MoHFW), Government of India (https://mohfw. gov.in) [18] and Worldometers (https://www.worldometers.info) [19] at daily scale.

3.2   Increase in Cases

The day to day cumulative increase in the positive cases are presented in Fig. 1 for all India and 10 states worst affected i.e. Maharashtra, Tamilnadu, Andhra Pradesh, Karnataka, Uttar Pradesh, Delhi, West Bengal, Bihar, Telangana and Assam. The analysis clearly shows the exponential growth of the COVID-19 spread in India and the states.

Figure 2 represents the cumulative increase in number of cases for top 10 states which have been ranked based on maximum number of cases till date for the pre-lockdown phase. As we observe from the graphs, Maharashtra, Delhi, and Telangana has a sharp increase in the number of cases. This could be attributed to the high population density and lack of social distancing from its citizens.

Similarly Fig. 3 represents the number of newly infected patients recorded per day for the given time period. When we analyze the increase in cases per day, we can see that Maharashtra the greatest number of cases per day recorded in the

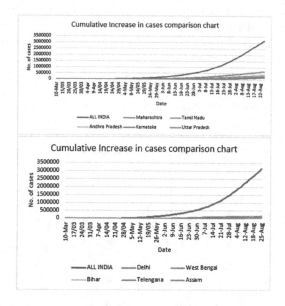

Fig. 1. The day to day increase of COVID-19 positive cases reported in India and 10 worst affected states

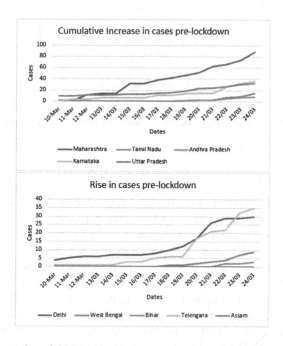

Fig. 2. The day to day COVID-19 positive cases reported during the pre-lockdown period in India and 10 worst affected states

country before lockdown was imposed by the government. Initially, Delhi and Telangana had a smaller number of cases per day then saw a drastic increase. Even Karnataka observed a steady rise in cases. Hence, lockdown had to be imposed to stem the rise in cases. Once the lockdown was lifted from 1st June, there has been a steady increase in cases. Maharashtra, Tamil Nadu, and Delhi continue to dominate the cases in India. West Bengal and Telangana have witnessed a major increase in the number of cases post lockdown (Fig. 4 and 5).

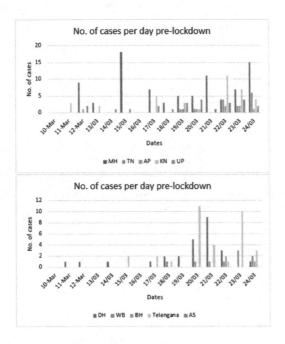

Fig. 3. Daily number of positive COVID-19 cases reported in India and 10 states during pre-lockdown period

3.3 Growth Rates

To get a measure of the severity of the spread, the growth rates of the COVID-19 cases are analyzed and presented in Fig. 6 and it is observed that Maharashtra and Telangana have a sharp increase in growth rates. These sharp increases represent a spike in cases. Similarly, this can be observed for West Bengal and Tamil Nadu towards the later stages and nearing towards the lockdown.

When we consider post lockdown (Fig. 7), we can see that Karnataka and Andhra Pradesh have prominent increases in growth rates, and the same carries for Assam and Telangana. Surprisingly, despite having the largest number of cases, Maharashtra, Delhi and Tamil Nadu maintain a consistent growth rate.

A deeper analysis from the above plots reveals that Maharashtra has a large spike in growth rate pre-lockdown whereas it maintains a rather consistent growth rate post lockdown. The reason is that, the state saw an increase

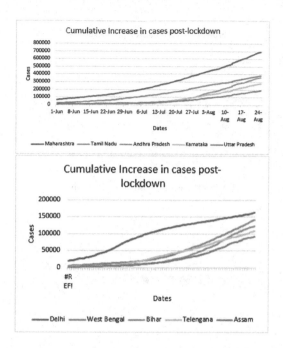

Fig. 4. The day to day increase of COVID-19 positive cases reported in India and 10 worst affected states during post-lock down period (i.e. June 01–24 August 2020)

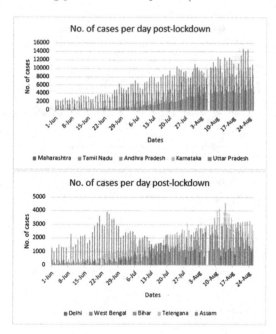

Fig. 5. Daily number of positive COVID-19 cases reported in India and 10 states during post-lockdown period (i.e. June 01–24 August 2020)

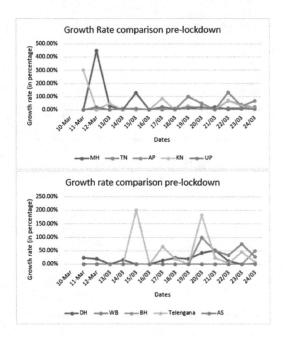

Fig. 6. Comparison of COVID-19 growth rate during pre-lockdown stage in India and states

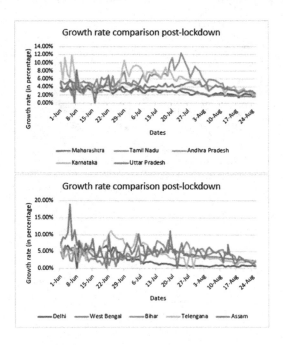

Fig. 7. Comparison of COVID-19 growth rate during post-lockdown stage in India and states

in newly infected people within a range of 0 to 20 in the pre-lockdown phase. Now for initial days of the pandemic, this was a huge increase as the number of cases was low. But when considering post lockdown phase, even though people are getting infected in the thousands, the number of cases is already high and hence the growth rate remains constant.

But this behavior is contrasting for the state of Assam. The state has a small growth rate during the pre-lockdown phase and hence the lesser number of cases. But when considering post lockdown phase, as the cases increase by the thousands, the graph shows major spikes in growth rate.

3.4   Impact of Lockdown

To assess the impact of lockdown in India the cases and growth rate during March-May 2020 is analysed and discussed. Figure 8 and 9 respectively represent the cumulative curve and the day to day reported cases in India and the states considered. As we can see from Fig. 8, Maharashtra and Delhi have a drastic increase in cases where as the remaining states have a rather consistent increase. This could be attributed to the amount of testing done as the increase in cases linearly dependent on the amount of testing done.

The above two graphs in Fig. 9 supplements to the above analysis during lockdown. Also, it was during this time that a lot of migrant workers were returning back to their hometowns and this could have contributed to the spread.

The growth rate during lockdown is presented in Fig. 10 and as per the analysis, Assam and Tamil Nadu saw a spike in cases where as Maharashtra and Delhi see a consistent growth rate despite having a large increase in cases. This could be attributed to the fact that Maharashtra and Delhi had a large number of cases going into lockdown whereas the remaining states had a rather smaller number. It can be attested that a smaller percentage increase in a large number is greater than a large percentage increase in a smaller number and hence the increase. For example, in the case of Maharashtra, on 12th March 2020, the cumulative number of cases rose from 2 to 11. That is a 450% increase in growth rate. Whereas, on 25th August 2020, the cumulative number of cases in Maharashtra rose from 6,82,383 to 6,93,398. 11,015 people were newly infected, but the growth rate for that day was only 1.61%.

4   Predictive Modelling

In this section, we develop and evaluate predictive models for forecasting the spread of virus for individual states in India using machine learning techniques. In particular, we have adopted time series analysis and deep learning approaches to predict the number of cases for a particular month based on the collective data of previous months.

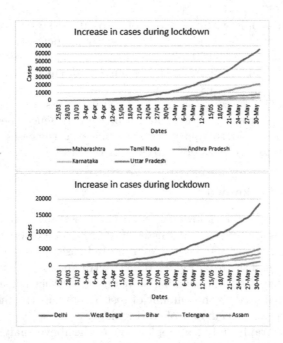

Fig. 8. The day to day increase of COVID-19 positive cases reported in India and 10 worst affected states during lockdown phase

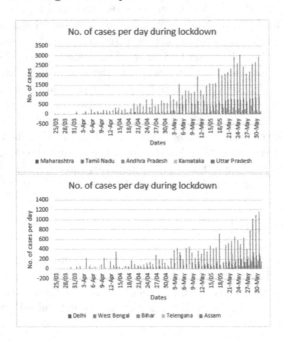

Fig. 9. Daily number of positive COVID-19 cases reported in India and 10 states during lock down phase

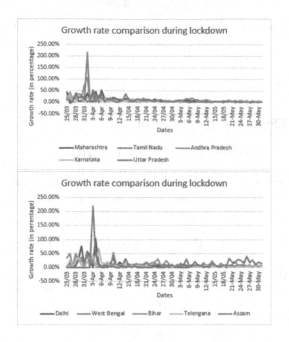

Fig. 10. Daily number of positive COVID-19 cases reported in India and 10 states during lock down phase

4.1 Time Series Analysis

We use ARIMA model to predict the rate of spread of COVID-19. There are two methods of prediction in ARIMA. The first method requires to differentiate the time series, apply auto-correlation function (ACF) and partial auto-correlation function (PACF) to determine d, q and p which represents the order of differencing, auto-regressive parameters and moving average parameter respectively. From the values obtained, ARIMA model is predicted to be ARIMA(p, d, q). One disadvantage in using this method is, the human error involved in deciding p and q from the graphs obtained by applying ACF and PACF. The predicted values differ significantly even with a small deviation in p and q. The second method uses an in-built function auto.arima where the model decides the perfect fit. Therefore, in this paper, auto.arima has been used. ARIMA works based on 'Backward-shift operator (B)', which back shifts the data by one time period. When applied to a time series Y and a period t, we get,

$$BY_t = BY_{t-1} \tag{1}$$

Multiplication by another B shifts the time period more by another time period,

$$B^2Y_t = B(BY_t) = BY_{t-1} = Y_{t-2} \tag{2}$$

For any time period n,

$$B^nY_t = Y_{t-n} \tag{3}$$

If y is the first difference of Y, then for time t,

$$y_t = Y_t - Y_{t-1} = Y_t - BY_t = B(1 - Y_t) \tag{4}$$

The differenced series y_t is obtained from the original time-series Y by multiplying by a factor of $(1 - B)$. Let z be defined as the first difference of y, i.e., z is the second difference of Y.

$$z_t = y_t - y_{t-1} = (1 - B)y_t = (1 - B)((1 - B)Y_t) = (1 - B)^2Y_t \tag{5}$$

Let us consider the equation for ARIMA (1, 1, 1) model.

$$y_t = Y_t - Y_{t-1} \tag{6}$$

$$y_t = \phi_1 y_{t-1} + e_t - \phi_1 e_{t-1} \tag{7}$$

where e_t is the random noise at time t. Using the backshift operator B,

$$y_t = (1 - B)Y_t \tag{8}$$

$$y_t = \phi_1 By_t + e_t - \theta_1 Be_t \tag{9}$$

$$(1 - \phi_1 B)y_t = (1 - \theta_1 B)e_t \tag{10}$$

Writing the equation in terms of Y,

$$(1 - \phi_1 B)(1 - B)Y_t = (1 - \theta_1 B)e_t \tag{11}$$

4.2 Deep Learning

We use LSTM algorithm to predict results. The data was modelled using Simple, Stacked and Convolutional LSTM. Out of these models, Stacked LSTM predicted the values with the least error rate, hence, their results have been included in the paper.

The mathematical equations for various gates and cells in an LSTM model are defined below,

$$
\begin{aligned}
i^t &= \sigma(W^t x^t + U^i h^{t-1}) &&\text{- Input Gate} \\
f^t &= \sigma(W^t x^t + U^f h^{t-1}) &&\text{- Forgot Gate} \\
o^t &= \sigma(W^t x^t + U^o h^{t-1}) &&\text{- Output/Exposure Gate} \\
\acute{c}^t &= tanh(W^c x^t + U^c h^{t-1}) &&\text{- New memory cell} \\
c^t &= f^t \acute{c}^{t-1} + i^t \acute{c}^t &&\text{- Final memory cell} \\
h^t &= o^t tanh(c^t)
\end{aligned}
$$

1. New memory generation - Input x^t and previous hidden state h^{t-1} generate a new memory $\tilde{c}^t$.
2. Input Gate - Uses input word and past hidden states to determine whether or not the input is worth preserving. It is used to gate the new memory. It produces i^t as an indicator for input gate.
3. Forget Gate - Similar to input gate except that, it does not make a determination of input word. It uses the input word and past hidden state to produce the forgot gate f^t.
4. Final memory generation - This step takes the advice of forget gate and forgets the past memory c^{t-1}. Similarly, it takes advice from the input gate i^t and gates the new memory. The sum of these two results produces the final memory c^t.
5. Output/Exposure Gate - This gate separates the final memory from the hidden state. Hidden states are used in every gate of LSTM and makes an assessment of the parts of the memory c^t that is to be exposed or present in the hidden state h^t. o^t is used to gate the point-wise tanh of the memory.

4.3 Evaluation of Predictive Models

The models have been evaluated and assessed for the top 10 COVID affected states, they are, Maharashtra, Tamil Nadu, Delhi, Andhra Pradesh, Karnataka, Uttar Pradesh, Gujarat, West Bengal, Telangana and Bihar. The number of cases for a particular month are predicted based on the collective data of all the previous months. In this paper, graphs of actual and predicted values and performance factors are presented for Maharashtra, Delhi and Karnataka for the months of April, August and November as shown in Figs. 11, 12, 13 and Table 1 respectively. Number of cases for the post-COVID period, i.e., the month of February, 2021 have been predicted using ARIMA, with the results recorded in Table 2. In the performance graphs, the red line indicates the actual number of cases. The blue portion represents the predicted values.

4.4 Model Validation

ARIMA and Stacked LSTM models have been used to predict the number of cases. It can be observed that the results obtained using Stacked LSTM have a wide deviation compared to the actual values, which are clearly visible in the loss graphs and performance factors. The results derived from ARIMA have high accuracy values, giving a reliable prediction.

The results obtained by training the model for March and predicting for April shows that the actual cases deviate exponentially compared to the predicted cases, hence, not effective.

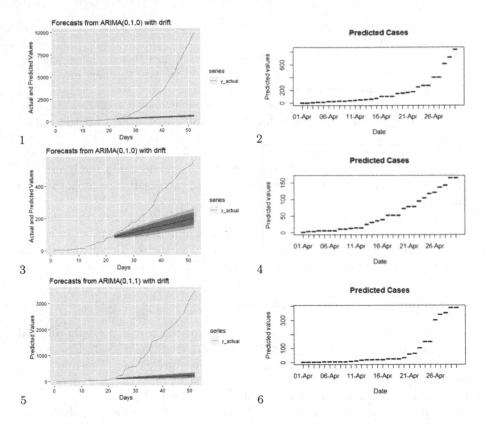

Fig. 11. Plots of actual and predicted values - Trained for March and predicted for April, (1) Maharashtra - ARIMA, (2) Mahrashtra - LSTM, (3) Karnataka - ARIMA, (4) Karnataka - LSTM, (5) Delhi - ARIMA, (6) Delhi - LSTM (Color figure online)

While training for March and April and predicting for May, the results obtained are slightly effective for the states of Andhra Pradesh, Telangana, Gujarat and Uttar Pradesh. However, the predicted values deviate from the actual values for the remaining states.

Training for March, April and May and predicting cases for June is applicable for Bihar, West Bengal, Gujarat, Maharashtra. The predicted values lie in the range of actual values. In the case of other states, the deviation between actual and predicted values has largely decreased and the predicted values are almost close to actual values.

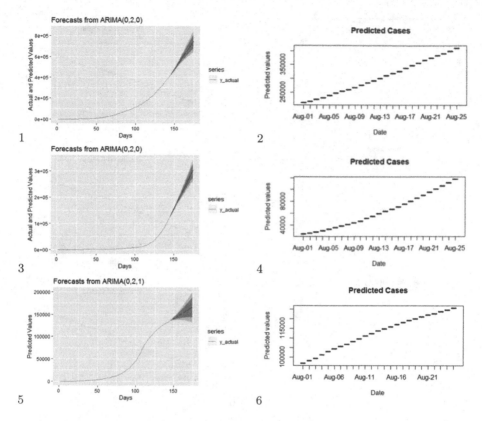

Fig. 12. Plots of actual and predicted values - Trained for March, April, May, June and July and predicted for August, (1) Maharashtra - ARIMA, (2) Mahrashtra - LSTM, (3) Karnataka - ARIMA, (4) Karnataka - LSTM, (5) Delhi - ARIMA, (6) Delhi - LSTM (Color figure online)

Training for March, April, May and June and predicting the cases in July is effective for Maharashtra, Tamil Nadu, Delhi, Karnataka, Andhra Pradesh, Gujarat and Telangana. Other states experience an exponential rise in the number of cases, due to which the model is not accurate.

Training for March, April, May, June and July and predicting for August works well for all the states. From the graphs, it is observable that the number of cases in Delhi is beginning to flatten.

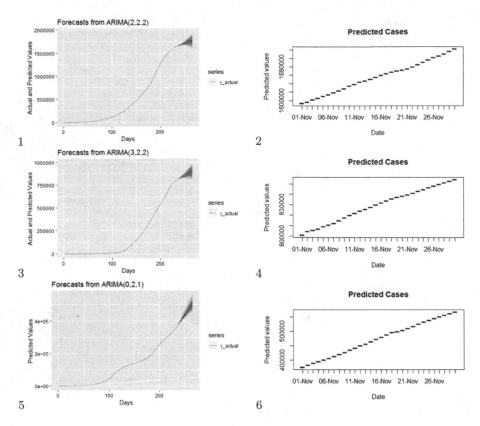

Fig. 13. Plots of actual and predicted values - Trained for March, April, May, June, July, August, September and October and predicted for November, (1) Maharashtra - ARIMA, (2) Mahrashtra - LSTM, (3) Karnataka - ARIMA, (4) Karnataka - LSTM, (5) Delhi - ARIMA, (6) Delhi - LSTM (Color figure online)

The performance values for the number of cases predicted form September onward shows that the model works well for all the states. It is observable that the curve starts to flatten. This forecast is clearly visible in the cases of Uttar Predesh, Gujarat and Bihar based on the predictions for February 2021.

From the performance values obtained, it is observable that, with the increase in the cumulative training months data, the model predicts with lower error rate and higher accuracy.

Table 1. Predicted value errors when compared to the actual values

Train months	Test month	Method	Performance (%)			
			Parameter	Maharashtra	Karnataka	Delhi
March	April	ARIMA	ME	3193.98	169.21	1392.10
			MAPE	77.02	47.24	83.35
			Accuracy	10.48	45.89	10.34
		LSTM	ME	3418.79	256.73	−1466.82
			MAPE	96.88	86.25	96.44
			Accuracy	3.87	17.17	5.61
March, April, May, June, July	August	ARIMA	ME	−5557.64	1935.76	440.04
			MAPE	1.20	2.17	0.36
			Accuracy	90.59	87.86	97.30
		LSTM	ME	245279.00	137538.50	−30675.10
			MAPE	44.93	70.26	20.77
			Accuracy	64.56	72.09	74.57
March, April, May, June, July, August, September, October	November	ARIMA	ME	16287.55	−4292.50	26336.63
			MAPE	0.93	−0.49	5.29
			Accuracy	99.06	99.50	94.36
		LSTM	ME	108022.00	613790.40	324995.40
			MAPE	195.01	315.22	213.57
			Accuracy	86.99	83.93	85.19

Table 2. COVID predicted value for February 2021

Date	Maharashtra	Tamil Nadu	Delhi	Andhra Pradesh	Karnataka	Uttar Pradesh	Gujarat	West Bengal	Telangana	Bihar
01/02	2032149	847914	646027	892348	942941	33500	271404	556984	300015	266611
02/02	2035382	848878	646712	892673	943690	33507	272253	556421	300441	267088
03/02	2038614	849842	647397	892998	944450	33514	273101	555821	300867	267566
04/02	2041847	850805	648082	893323	945218	33522	273949	555187	301294	268043
05/02	2045080	851769	648767	893647	945987	33529	274798	554520	301720	268520
06/02	2048313	852733	649452	893972	946751	33537	275646	553819	302147	268998
07/02	2051546	853697	650137	894297	947504	33544	276494	553086	302573	269475
08/02	2054779	854660	650822	894622	948252	33551	277343	552321	303000	269952
09/02	2058012	855624	651506	894947	949003	33559	278191	551526	303426	270430
10/02	2061245	856588	652191	895272	949762	33566	279039	550699	303852	270907
11/02	2064478	857552	652876	895597	950529	33573	279888	549843	304279	271384
12/02	2067710	858515	653561	895922	951297	33581	280736	548958	304705	271861
13/02	2070943	859479	654246	896247	952060	33588	281584	548045	305132	272339
14/02	2074176	860443	654931	896572	952814	33595	282433	547103	305558	272816
15/02	2077409	861407	655616	896896	953563	33603	283281	546135	305984	273293
16/02	2080642	862370	656301	897221	954315	33610	284129	545139	306411	273771
17/02	2083875	863334	656986	897546	955073	33618	284978	544118	306837	274248
18/02	2087108	864298	657671	897871	955839	33625	285826	543071	307264	274725
19/02	2090341	865262	658355	898196	956607	33632	286674	541999	307690	275201
20/02	2093574	866225	659040	898521	957370	33640	287523	540902	308116	275680
21/02	2096806	867189	659725	898846	958125	33647	288371	539781	308543	276157
22/02	2100039	868153	660410	899171	958875	33654	289219	538637	308969	276635
23/02	2103272	869117	661095	899496	959626	33662	290068	537470	309396	277112
24/02	2106505	870080	661780	899821	960385	33669	290916	536281	309822	277589
25/02	2109738	871044	662465	900145	961150	33676	291764	535069	310249	278067
26/02	2112971	872008	663150	900470	961917	33684	292612	533836	310675	278543
27/02	2116204	872972	663835	900795	962680	33691	293461	532582	311101	279021
28/02	2119437	873936	664520	901120	963435	33698	294309	531307	311528	279498

5   Conclusion

In this paper, we analyzed the impact of COVID 19 for the individual states in India in a comparative manner. In particular, we developed a dataset containing the number of COVID-19 cases for India and conducted exploratory data analysis to analyze the growth rates and further examined the effectiveness of the lockdown on the growth rates for the states most affected by the virus. Finally, we developed predictive models to forecast the spread of virus for the states most affected by virus using time series analysis and deep learning and evaluated them in a comparative manner. Our models predict the cases with 97.30% accuracy, hence, can be used to guide the development of mechanisms for optimal resource allocation of healthcare systems and response.

References

1. Gorbalenya, A., et al.: The species Severe acute respiratory syndrome-related coronavirus: classifying 2019-nCoV and naming it SARS-CoV-2. Nat. Microbiol. **5** (2020). https://doi.org/10.1038/s41564-020-0695-z
2. Is the coronavirus airborne? Experts can't agree. https://www.nature.com/articles/d41586-020-00974-w
3. Li, Q., et al.: Early transmission dynamics in Wuhan, China, of novel coronavirus-infected pneumonia. New Engl. J. Med. **382** (2020). https://doi.org/10.1056/NEJMoa2001316
4. Morawska, L., Cao, J.: Airborne transmission of SARS-CoV-2: the world should face the reality. Environ. Int. **139** (2020). https://doi.org/10.1016/j.envint.2020.105730
5. Coronavirus Outbreak in India. http://www.covid19india.org/
6. Killerby, M., et al.: Human coronavirus circulation in the United States 2014–2017. J. Clin. Virol. **101** (2018). https://doi.org/10.1016/j.jcv.2018.01.019
7. Aarogya Setu: MyGov (2020). https://www.mygov.in/aarogya-setu-app/
8. Ahmed, N., et al.: A survey of COVID-19 contact tracing apps. IEEE Access **8**, 134577–134601 (2020). https://doi.org/10.1109/ACCESS.2020.3010226
9. Kissler, S., Tedijanto, C., Goldstein, E., Grad, Y., Lipsitch, M.: Projecting the transmission dynamics of SARS-CoV-2 through the post-pandemic period (2020). https://doi.org/10.1101/2020.03.04.20031112
10. Chen, Y.-C., Lu, P.-E., Chang, C.-S., Liu, T.-H.: A time-dependent SIR model for COVID-19 with undetectable infected persons. IEEE Trans. Netw. Sci. Eng. **7**(4), 3279–3294 (2020)
11. Giamberardino, P., Iacoviello, D.: Optimal resource allocation to reduce an epidemic spread and its complication. Information **10**, 213 (2019). https://doi.org/10.3390/info10060213
12. Li, H., Bailey, A., Huynh, D., Chan, J.: YouTube as a source of information on COVID-19: a pandemic of misinformation? BMJ Glob. Health **5** (2020). https://doi.org/10.1136/bmjgh-2020-002604
13. Jin, F., Dougherty, E., Saraf, P., Cao, Y., Ramakrishnan, N.: Epidemiological modeling of news and rumors on Twitter. In: Proceedings of the 7th Workshop on Social Network Mining and Analysis (2013). https://doi.org/10.1145/2501025.2501027

14. Rustam, F., et al.: COVID-19 future forecasting using supervised machine learning models. IEEE Access **8**, 101489–101499 (2020). https://doi.org/10.1109/ACCESS.2020.2997311

15. Jamshidi, M., et al.: Artificial intelligence and COVID-19: deep learning approaches for diagnosis and treatment. IEEE Access **8**, 109581–109595 (2020). https://doi.org/10.1109/ACCESS.2020.3001973

16. Hussain, A.A., Bouachir, O., Al-Turjman, F., Aloqaily, M.: AI techniques for COVID-19. IEEE Access **8**, 128776–128795 (2020). https://doi.org/10.1109/ACCESS.2020.3007939

17. Shoeibi, A., et al.: Automated detection and forecasting of COVID-19 using deep learning techniques: a review (2020)

18. Ministry of Health and Family Welfare, Government of India. https://www.mohfw.gov.in/

19. COVID-19 Coronavirus Pandemic, Worldometer. https://www.worldometers.info/coronavirus/#countries

ASIF: An Internal Representation Suitable for Program Transformation and Parallel Conversion

Sesha Kalyur$^{(\boxtimes)}$ and G. S. Nagaraja

Department of Computer Science and Engineering, R. V. College of Engineering, VTU, Bangalore, India
nagarajags@rvce.edu.in

Abstract. Adding the Transformation and Parallelization capabilities to a compiler requires selecting a suitable language for representing the given program internally. The higher level language used to develop the code is an obvious choice but supporting the transformations at that level would require major rework to support other higher languages. The other choice is to use the assembly representation of the given program for implementing transformations. But this would require rework when supporting multiple targets. These considerations lead to the development of an internal representation that is not tied to any specific higher level language or hardware architecture. However, creating a new Internal Representation for a compiler that ultimately determines the quality and the capabilities of the compiler offers challenges of its own. Here we explore the design choices that determine the flavor of a representation, and propose a representation that includes, Instructions and Annotations that together effectively represent a given program internally. The instruction set has operators that most resemble a Reduced Instruction Set architecture format, and use three explicit memory operands which are sufficient for translation purposes and also simplify Symbolic Analysis. In addition to instructions, we support Annotations which carry additional information about the given program in the form of Keyword-Value pairs. Together instructions and annotations contain all the information necessary to support Analysis, Transformation and Parallel Conversion processes. ASIF which stands for Asterix Intermediate Format, at the time of writing is comparable to the cutting-edge-solutions offered by the competition and in many instances such as suitability for Program Analysis is superior.

Keywords: Optimization · Transformation · Parallelization · Internal representation · IR · Three address format · Instruction set architecture · ISA

© ICST Institute for Computer Sciences, Social Informatics and Telecommunications Engineering 2021
Published by Springer Nature Switzerland AG 2021. All Rights Reserved
N. Kumar et al. (Eds.): UBICNET 2021, LNICST 383, pp. 150–168, 2021.
https://doi.org/10.1007/978-3-030-79276-3_12

1   Introduction

The Internal Representation (IR) of an optimizing and parallelizing compiler, plays a very important role in the overall compiler chain. The suitability of the IR for Translation, Transformation and Parallel Conversion ultimately determines the success and popularity of the optimizing compiler.

What are the available choices? At one end of the spectrum we have the source code in some high level language, and at the other end lies the binary object code for some hardware architecture. Both formats have their own advantages and disadvantages. For instance, working on the source code can result in simple solutions, but would require rework for supporting multiple languages. A solution built on top of the binary, has the advantage of supporting a solution at program run time. The solution is also inherently complex since much of the higher level information is missing. Rework is required to support binaries for additional hardware targets. To overcome these limitations, the source code in some high level language is first translated to an intermediate format. Optimizing transformations are carried out on the intermediate code. Then the transformed code is translated to object code for the chosen hardware architecture. It is easy to see why the choice of a suitable intermediate format assumes utmost importance in the transformation and parallelization process.

Let us next look at the various choices that exist for structuring IR. The answer lies in the different Instruction Set Architecture (ISA) choices that are available. In a Stack Based ISA, all operands exist on a stack which is basically a last-in first-out buffer. In an Accumulator based architecture, one of the operands is implicitly provided from, a special register called the Accumulator. Register based ISA are characterized by operands, resident in programmable general purpose registers. When all operands of all instructions are held in memory locations, the resulting architecture is said to be Memory based. An ISA is hybrid in nature, when the source and destination of the operands, can be memory or registers and is instruction specific [1].

Instruction set architectures (ISA), can also be classified according to the number of explicit operands, that are present in an instruction. In a two address ISA one of the operands also serves as the implicit result. On the other hand, in a three address ISA all the source and the result operands are explicit and live in registers or memory locations.

ISAs can also be classified based on style, as Reduced Instruction Set Computer (RISC) architectures and Complex Instruction Set Computer (CISC) architectures. In RISC, the only operations that take memory operands, are the Load and Store instructions. However in a CISC architecture, instruction formats are more complex and basically any instruction can operate on memory operands.

Typically, instructions in an ISA, can belong to one of the following categories namely,

- Arithmetic (Addition and Subtraction),
- Logical (And and Or),

- Data transfer (Load and Store),
- Control (Jump, Conditional Jump and Procedure Call and Return),
- System (System Call),
- Floating Point (Floating point Multiply),
- Decimal (Decimal Divide),
- String (String Move),
- Graphics (Compression and Decompression).

The sections that follow provide more clarity and exposure to our research work and the findings. Section 2 looks at the previous research work in this area. Section 3 discusses in depth the features of ASIF. Section 4 closely looks at ASIF as it is serves many transformation and parallel conversion needs. Section 5 is the Analysis section where ASIF is analyzed and compared with the competition. Section 6 is the Results section where ASIF results is compared with the results of two other open source technologies. Conclusion section is the last section in the paper, where we highlight the contributions of this research work and briefly look at the future.

2   Related Work

Search of the published literature provides interesting results. Polaris IR was developed to analyze and optimize FORTRAN programs, based on the abstract syntax tree concept [2]. HTZ IR is based on the Hierarchical Task Graphs, which allows efficient manipulation of higher level language programs, for analysis and visualization purposes [3]. The SUIF compiler suite is unique in the sense, that it supports an extensible internal representation framework, that allows users to augment the instruction set, via grammar specifications to support new analysis techniques [4,5]. A few others use internal representations at multiple levels, with an instruction based representation at the mid-level [6]. Other internal representations include those based on the Polyhedral model [7–9], Linear algebra [10] and Symbolic algebra [11].

There has been a resurgence of research interest in the field of Compiler Optimization and Parallelization and a search of the literature yields the following results. Promis compiler employs Hierarchical Task Graphs as IR at three levels, exploits dynamic, loop and instruction level parallelism present in the program [12]. IBM JIT compiler uses Java Byte Code as its IR and runs its analysis and optimization steps such as, Common Sub-expression Elimination among others on the byte codes [13]. Polly compiler uses LLVM-IR as the IR and coupled with a Polyhedral model, performs low-level optimizations such as converting to OpenMP enabled code [14]. Emscripten compiler translates LLVM programs to Javascript and uses the input LLVM program as IR and performs optimizations such as Variable Nativization and Re-looping on LLVM-IR [15].

Rose compiler supports source to source transformations, generating OpenMP and MPI enabled codes, and uses an IR that is structured as an Abstract Syntax Tree [16,17]. Cetus is a source to source infrastructure that uses a flat IR representation which is designed as a Java Class hierarchy, powerful enough to represent statements and expressions in the language and is able to generate OpenMP enabled code among others. [18,19] SPARK is a VHDL synthesis compiler that uses Hierarchical Task Graphs as IR and implements transformations such as Common Sub-expression Elimination [20].

A search of the state-of-the-art in industry reveals some interesting candidates as well. We have Low Level Virtual Machine (LLVM) with an IR, that supports mid-level analysis and transformations [21,22]. The IR in LLVM is structured in the Reduced Instruction Set Computer (RISC) format, and supports a simple instruction set that includes Add, Subtract etc. Among the competition, LLVM comes closest to ASIF in structure and functionality.

The internal representation in the Open64 compiler is called WHIRL, which can exist at multiple levels [23]. The highest form has a tree representation, the mid level has a graphical form and the lowest form is in the form of expressions and statements ideal for performing optimizations.

GCC supports a high level tree based internal representation called GENERIC, and a three address representation called GIMPLE, which is especially useful for optimization analysis such as loop analysis [24]. Intel uses a LLVM based internal representation [25,26]. Microsoft Visual Studio compiler uses an IR based on the syntax tree model [27,28]. Path-scale EKOPath is based on the Open64 compiler infrastructure, and hence uses WHIRL as the internal representation [29,30].

After careful examination of the existing IR solutions, we concluded that an ideal IR representation would support the following:

- Efficient Symbolic Processing
- Efficient Flow Analysis
- Efficient Dependence Analysis
- Efficient Loop Analysis
- Efficient Array Analysis
- Efficient Alias Analysis

However existing IR solutions though feature rich, fell short of meeting all our requirements and hence we decided to pursue design of the ideal IR solution that meets our expectations. The result of this research and design is ASIF, the IR solution we are proposing for our compiler. ASIF is a three-address ISA, that explicitly names the result, and all the source operands. All operands are memory resident, except for a few that take immediate operands and labels and a few operations such as Load and Store. The decision to eliminate register operands and use explicit result and source operands, has helped us to simplify symbolic processing and analysis. There will be a separate Register Allocation phase later on in the Asterix chain, where memory operands are replaced by registers prior to processing. ASIF will be complemented by other graph based and algebraic IR formats, during flow and dependence analysis. The following sections will provide more details and rationale, for selecting ASIF as our ideal IR solution.

3   ASIF

ASIF is a short form for ASterix Intermediate Format. *Asterix Intermediate Format (ASIF)* is a representation where instructions in general follow the three address format, with explicit Result and two Source operands. All operands are memory operands (no register operands) with the exception of Labels, Immediate operands, Load and Store instructions. ASIF also supports a number of annotations which emit additional information pertaining to Loops, Control, Function and Structure elements among others which are used for analysis purposes. Also since we only use memory operands, some of the symbolic analysis phases such as Dependence and Flow analysis becomes relatively simple. The following subsections will provide additional details about selected instructions and annotations.

3.1 Asterix Transformation and Parallelization Pipeline

Asterix our compiler is designed to handle auto-parallelization of C programs and perform transformations that support parallel conversions. ASIF-IR provides the backbone on which the various phases of Asterix are structured [31–36].

The phases involved in Asterix include Translation, Analysis, Transformation, Parallelization, Code Generation, Orchestration, Distribution and Execution. Figure 1 provides a pictorial view of how the phases are organized in Asterix.

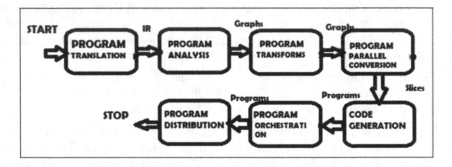

Fig. 1. Program transformation and parallelization pipeline in asterix

3.2 Rationale for ASIF

Several internal representations or models, have been used in the domain of program transformation and parallelization [37,38]. For instance, an internal representation based on the Parse tree, is useful for evaluating expressions and some transformations, such as Common sub-expression elimination [39]. Graph based representations have been useful in implementing, several optimizations and parallelization and are especially useful for visualization purposes [40–42]. Polyhedral representations have found use in Vectorization and Parallelization [7–9]. Representations based on Linear algebra, are useful for carrying out certain linear optimizations, involving arrays such as Data layout transformations

[10]. Symbolic algebra techniques are especially useful, for carrying out transformations involving loops and arrays [11].

Translation of the source code, directly to an internal representation based on trees, graphs and algebra has certain advantages such as efficiency. However using multiple models for transformation and parallelization, would require multiple parsers, one each for each source language and representation. Translating to an internal representation such as a three address form and then generating the appropriate models, reduces the complexity by a great extent.

ASIF our internal representation, which is a three address based instruction format, is first generated after parsing the source program. Subsequently other internal representations based on graphs and SSA, are generated as needed from the ASIF program. This puts a check on the combinatorial explosion alluded to earlier. Majority of the instructions in ASIF operate on memory based operands, as opposed to register resident operands, which eases the Flow and Dependence analysis steps especially the Symbolic processing part. Since all argument and result operands for most of the ASIF instructions are explicitly specified, analysis process gets simplified. After the transformation and parallelization phases are completed, ASIF code is translated to the appropriate code, in the assembly language of the target architecture.

3.3   ASIF Instruction Formats

ASIF is a complete instruction set architecture (albeit for a virtual machine), in the sense that it includes instructions from several categories, such as Arithmetic, Logical, Data Transfer, Jumps, Branches, Call and Return, Floating point, Scalar and Predicated among others. Since it is not possible to include every type of instruction defined in ASIF due to limited space only a selected few from each set is chosen for further discussion below.

The Arithmetic Set includes the instructions ADD, SUB, MUL, DIV, REM among others with usual semantics, The Idiom Set includes, INC, DEC, INK, DEK among others. The Compare Set includes LTH, LTE, GTH, GTE, EQL, NEQ as representatives. The Logical Set has the instructions AND, LOR, NOT, XOR. The Bitwise Set is representaed by BAN, BOR, BNO, BXR, ROR, ROL. The Memory Set includes MRD, MWR, ADR, PTR as representatives. Jump and Branch Set is represented by JMP, JMT, JMF, JNE, JEQ, BRC, BRT, BRF, BNE, BEQ. Function and Stack Set includes DEF, FNC, FNR, PSH, POP among others. Floating Point Set has the following instructions, FADD, FSUB, FMUL, FDIV. The Vectorization Set includes VADD, VSUB, VMUL, VDIV among others. The Predication Set has the instructions PADD, PSUB, PMUL, PDIV as representatives.

It should be noted that the ASIF instructions are necessary and sufficient for code generation purposes. However to support complex analysis for transformation and parallel conversion process, code annotations which carry extra information are necessary and are discussed next.

3.4   ASIF Annotation Formats

In addition to emitting ASIF code, the front end of our compiler, also outputs annotations, which are not executable content, but play a very important part in the transformation and parallelization phases, by passing useful information, to the optimization and back-end phases. In a broad sense, annotations are of two main types namely, those that target code and those that target data. It is in a sense, convenient to view them as Pseudo-instructions.

When translating an imperative high level language, such as C, code annotations could be used to pass additional information about modules, functions, function-calls, function-returns, loops, loop-nests, conditions, condition-nests, recursion, expressions and so on. In a similar vein, when data structures and variables are defined and accessed, additional information gleaned at translation time, can be passed on with the help of annotations. Annotations are organized as a list of name-value pairs which are processed downstream. For instance we could ascertain, if the concerned variable is a structure, union, bit-field, array or pointer and specific details about the type of the data structure. When new data types are defined using type-def, additional information about the new type, can also be passed on, with the help of annotations.

Annotations are of two categories namely Code and Data. In general all annotations contain information on the span of lines where the element is found in the source file and also the span of instruction numbers where you can find the element. In addition element specific name-value pairs contain information specific to the element. Some of the code annotations include the ones for the Module, Function Definition, Function Call, Function Return, Loop, Loop-Nest, Control, Control-Nest, Function Recursion and Expression. Similarly Data Annotations include the ones for Type Definition, Structure Definition, Union Definition, Bitfield Definition, Enum Definition, Variable Definition, Array Definition and Pointer Definition. For instance Function Definition annotation would include information about parameters passed, their names and types. Similarly the Annotation for a Structure Definition would include information about the fields, their names and types. Other annotations are similar in their working and they carry information specific to the element.

It should be noted that the list of annotations presented earlier, is not meant to be exhaustive, but only illustrative. As seen most of these annotations, focus on the definition part of a particular code or data element. However other annotations are possible, such as at the data access points for instance, which could be defined along similar lines.

4   ASIF as an Enabler of Transformation and Parallelization

The IR instructions generated as a result of translating high level source code are necessary and sufficient to generate the object code. The Operator set required for translating expressions in some high level language is finite and once defined are sufficient to generate machine code. IR also includes code features for identifying basic blocks, loops, conditions, functions, function calls and returns which are handled with labels as instruction targets and use of conditional and non-conditional jumps or branches. For machine defined and user defined data items, both scalar and array items can be defined knowing the size and alignment restrictions.

However for program analysis, transformation and parallel conversion we need to glean additional information from the source. How do we pass on this information downstream? Information gathered is formatted as annotations which are specific for each kind of program construct which will be used by analysis passes further downstream. In the absence of which, we would have to share binary information between the front line translation phases and the downstream optimization and code generation phases, which can create serious version management issues between phases.

We shall next look at how the information contained in the ASIF instructions and annotations are helpful to carryout the various analysis and code transformations.

4.1 Code Motion

Code Motion is an optimization opportunity that involves a loop, the induction variables and an expression that does not involve the index or induction variables of the loop. See Listing 1.1 on page 8 for a Code Motion example in C.

Listing 1.1. Code Motion Example in C

```
10: while ( i <= 2 * limit − limit / 2)
11:    i++;
```

In the code above the loop induction variable is compared to an expression whose value does not change with each iteration of the loop. See Listing 1.2 on page 8 for a Code Motion example in ASIF.

Listing 1.2. Code Motion Example in ASIF

```
0: $$label0 :
1:    MUL $T0, 2, limit
2:    DIV $T1, limit , 2
3:    SUB $T3, $T0, $T1
4:    INC i
5:    JLE i , $T3, $$label0
6: $$label1 :
```

The expression is easy to spot in C code. However in the translated ASIF code, the expression is spread over several instructions and hence not easy to spot. So an annotation identifying the expression is written to simplify identifying the expression. See Listing 1.3 on page 8 for annotations for the Code Motion example.

Listing 1.3. Code Motion Annotations

```
ANO {category :: code, kind :: expression,
identifier :: 'example(1).function(1).expression(1)',
symbolic :: expr_1, file :: example1.c, span ::10:10,
spread ::1−3 operators ::[MUL, SUB, DIV],
operands ::[D2, $limit, $limit, D2]}

ANO {category :: code, kind :: expression,
identifier :: 'example(1).function(1).expression(2)',
symbolic :: expr_2, file :: example1.c, span ::10:10,
spread ::1−4 operators ::[LTE], operands ::[$i, exp_1]}

ANO {category :: code, kind :: loop, class :: while,
identifier :: 'example(1).function(1).loop(1)'
symbolic :: loop_1, file :: example1.c, span ::10:11,
spread ::0−6 conditions ::[expr_2],
inductions ::[i], start ::, end ::, step ::, block ::[5]
```

Three annotations are written out for the given example. the first involves the expression that is a loop invariant and can be moved above and the second expression that serves as the loop condition plus the annotation for the loop itself. Since expression 1 involves operands that are not changed every loop iteration (involves no induction variables), it will be easy to identify it as a loop invariant and so the instructions, 1 to 3 will be moved above the loop and the expression value moved to a temporary and will be substituted for the expression in the loop expression as seen below. See Listing 1.4 on page 9 for annotations for the Code Motion example in ASIF optimized.

Listing 1.4. Code Motion in ASIF (Optimized)

```
0:   MUL $T0, 2, limit
1:   DIV $T1, limit, 2
2:   SUB $T3, $T0, $T1
3: $$label0 :
4:   INC i
5:   JLE i, $T3, $$label0
6: $$label1 :
```

4.2 Data Dependence

Data Dependence is an analysis step where dependence between instructions in a strided section of a program is searched. Since ASIF employs explicit memory operands, this step becomes simple. See Listing 1.5 on page 9 for the C version of the Data Dependence example. As seen in line 2, B[i] depends on the value of A[i] set in the previous statement.

Listing 1.5. Data Dependence Example in C

```
1:   A[i] = 10;
2:   B[i] = A[i] * i;
```

See Listing 1.6 on page 10 for the ASIF version. As seen, the dependence arises in instruction 2 from the fact that the result operand is dependent on the result operand A[i] set in the previous line. For simple Data Dependence, you do not need annotations to be written out and just the ASIF instructions would suffice.

Listing 1.6. Data Dependence Example in ASIF

```
1:   MOV A[i], 10
2:   MUL B[i], A[i], i
```

4.3   Loop Dependence

Loop Dependence is an analysis step where dependence that exists between iterations of the loop is searched. See Listing 1.7 on page 10 for the C version of the Loop Dependence example. We see that each iteration is dependent on its previous iteration.

Listing 1.7. Loop Dependence Example in C

```
1:   A[0] = 1;
2:   for (i = 1; i < N; i++)
3:     A[i] = A[i-1] * i;
```

See Listing 1.8 on page 10 for the ASIF version of the Loop Dependence example. As seen most of the information present in the high level code is missing in ASIF.

Listing 1.8. Loop Dependence Example in ASIF

```
1:   MOV A[0], 1
2:   MOV i, 1
3: $$label0:
4:   MUL A[i], A[i-1], i
5:   INC i
6:   JLT i, N, $$label0
7: $$label1:
```

See Listing 1.9 on page 10 for the annotations for the Loop Dependence example. We see it has necessary information to complete loop dependence analysis. The loop block is instruction 4 which after simple analysis we can say that there is loop dependence which can prevent many optimizations including loop parallelization.

Listing 1.9. Loop Dependence Example Annotations

```
ANO {category :: code, kind :: loop, class :: for,
identifier :: 'example(3).function(3).loop(3)'
symbolic :: loop_3, file :: example3.c, span :: 1:3,
spread :: 1 − 7, conditions :: , inductions :: [ i ],
start :: 0, end :: N, step :: 1, block :: [4] }
```

4.4 If Conversion

If Conversion or *Predicated Execution* is a transformation that converts an conditional to strided code. See Listing 1.10 on page 11 for the C version of the If Conversion example.

Listing 1.10. If Conversion Example in C

```
1:   if (A == 0)
2:     B = C + D
```

See Listing 1.11 on page 11 for the ASIF version of the If Conversion example.

Listing 1.11. If Conversion Example in ASIF

```
1:  JNE $$label1 , A, 0
2:  ADD B, C, D
3: $$label1 :
```

See Listing 1.12 on page 11 for the annotations for the If Conversion example.

Listing 1.12. If Conversion Example Annotations

```
ANO {category :: code, kind :: expression,
identifier :: 'example(3).function(3).expression(2)',
symbolic :: expr_2, file :: example3.c, span :: 1 − 1,
spread :: 1 − 1 operators :: [EQL], operands :: [$A, 0] }
```

```
ANO {category :: code, kind :: control, class :: if,
identifier :: 'example(3).function(3).control(3)'
symbolic :: control_3, file :: example3.c, span :: 1:2,
spread :: 1 − 3, start :: 0, end :: 5, step :: 1, block :: [2] }
if_condition :: expr_2, else_if_conditions :: ,
else_condition :: , if_block :: [2],
else_if_block :: , else_block :: }
```

As seen from the annotation, we see that the concerned element is a conditional and the condition for the element is written out as an expression. The body of the instruction is instruction 2 which is a simple Add instruction. As seen below the conditional is replaced by the two Predicated instructions. The information in the annotation is sufficient to produce the transformed ASIF code. See Listing 1.13 on page 11 for the optimized version of ASIF for the If Conversion example.

Listing 1.13. If Conversion Example ASIF (Optimized)

```
PEQL  A,  0,  B
PADD  B,  C,  D
```

4.5   Vectorization

Vectorization is an optimization in which a scalar operation on an array entry, in a loop is replaced by its corresponding Vector operation and in the process eliminating the loop.

See Listing 1.14 on page 12 for the C version of the Vectorization example. We see that each array entry is being multiplied by a constant and clearly a Vectorization opportunity.

Listing 1.14. Loop Vectorization Example in C

```
for  (i  =  0;  i  <  N;  i++)
  A[i]  =  A[i]  *  K;
```

See Listing 1.15 on page 12 for the ASIF version of the Vectorization example. The details available in the C program is hidden here and it would require extensive analysis to spot the Vectorization opportunity.

Listing 1.15. Loop Vectorization Example in ASIF

```
1:  $$label0 :
2:     MUL A[i] , A[i] , K
3:     INC i
4:     LTH $$label0 , i , N,
5:  $$label1 :
```

See Listing 1.16 on page 12 for the annotations for the Vectorization example. From the annotation, we see that instruction 2 is a simple multiplication with a constant and for all iterations it remains the same. There is no cross iteration dependence as seen from the instruction. It is easy to conclude that the loop is Vectorizable.

Listing 1.16. Loop Vectorization Example Annotation

```
ANO {category :: code , kind :: loop , class :: for ,
  identifier :: 'example(3) . function (3) . loop(3) '
  symbolic :: loop_3 , file :: example3 . c , span :: 1 : 3 ,
  spread :: 1 − 7 , conditions :: , inductions :: [ i ] ,
  start :: 0 , end :: N, step :: 1 , block :: [ 2 ] }
```

See Listing 1.17 on page 12 for the optimized version for the Vectorization example. The loop is eliminated by Vectorization.

Listing 1.17. Loop Vectorization Example in ASIF (Optimized)

```
2:     VMUL A, N, K
```

What was shown in the form of example transformations is only a small fraction of what is achievable with ASIF instructions and annotations working together. Some of the other application areas include but not-limited to, *Alias Analysis, Machine Idiom Optimization, Procedure Sorting, Reduction In Strength, Algebraic Simplification, Copy Propagation* and *Constant Propagation*.

5   Analysis

From a analysis perspective, we compared ASIF with the state-of-the-art solutions and captured their suitability for use as IR especially, with a focus on their ability to support transformation and parallel conversions. See Table 1 for comparision of ASIF with the competition. From the comparision it is obvious that ASIF matches the competition in most respects and ends up being superior in some cases especially its support for transformation and parallel conversion.

6   Results

We wanted to compare the performance of Asterix/ASIF our compiler with Gcc/Gimple and Clang/LLVM for three programs written in C namely do_all.c, do_depend.c and do_across.c. do_all.c has a simple loop with no data or loop carried dependencies. do_depend.c contains a similar loop with one data dependence for every iteration of the loop. Finally, do_across.c has one loop carried dependence for every iteration. Specifically each iteration depends on the previous one.

The results of the measurement are captured in the table below. See Table 2 for details.

Both times and speedups were plotted for each case and the plots are given below. Figure 2 provides a plot of the times taken by the three programs to run. Figure 3 provides a plot of the speedups exhibited by the three programs.

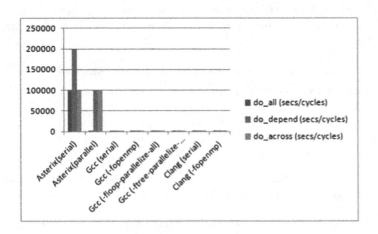

Fig. 2. Program run times

Table 1. IR competitive analysis

No.	Feature	ASIF	LLVM	GIMPLE	Open64	MASM	NASM
1	Specific name	ASIF	LLVM	GIMPLE	WHIRL	MASM	NASM
2	Levels	None	None	High and Low	Very high, High, Mid, Low, Very low	None	None
3	Data type support	Entire C set supported	Extensive	Moderate	Extensive	Moderate	Moderate
4	Structured control flow statements	Through annotations	Using Metadata	None	Elaborate support	None	None
5	Instruction Format	3 address	3 address	3 address	N/A	2 address	2 address
6	Operand type	Memory and immediate	Memory, Register and constants	Memory, Register and constants	Memory, Register and Constants	Memory, Register and Constants	Memory, Register and Constants
7	Format of Data for analysis	Annotation	Metadata	Annotation	Structured control flow statements	None	None
8	Instruction specialization	No condition codes	Condition codes used	None	None	None	None
9	Floating point opcodes	Yes	Yes	No	No	None	None
10	Vector specific opcodes	Yes	No (via CC)	No	No	No	No
11	Predicated opcodes	Yes	No support for predication	No	No	No	No
12	Analysis data access	Name, value pairs	Name, value pairs	Query functions to support analysis	N/A	None	None
13	Intrinsic support	None	Extensive support	None	Supported	Supported	None
14	Transformation support	High	High	Moderate	Moderate	None	None
15	Parallelization support	High	High	Moderate	None	None	None

While the times gathered for Gcc/Gimple amd Clang/LLVM are a little inconclusive, the cycles reported by Asterix/ASIF is more realistic. Asterix/ASIF has correctly reported that speedup gained by parallelization is equal to the number of iterations in the loop for the do_all program since there no dependencies of any kind. For do_depend it has reported correctly that for every iteration which has two instructions, there is one dependence. That is the reason why the

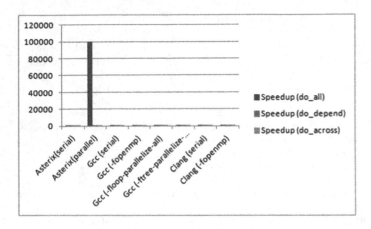

Fig. 3. Program speedups

Table 2. Performance results

No.	Compiler/ Flags	do_all sec/cyc	do_depend sec/cyc	do_across sec/cyc	do_all(R) ratio	do_depend ratio	do_across ratio
1	Asterix (serial)	100000	200000	100000	N.A	N.A	N.A
2	Asterix (parallel)	1	100000	100000	100000	2	1
3	Gcc (serial)	1.53	1.609	1.89	N.A	N.A	N.A
4	Gcc (-fopenmp)	1.53	1.608	1.828	1	1	1.034
5	Gcc (-floop-parallelize-all)	1.515	1.593	1.875	1.009	1.01	1.008
6	Gcc (-ftree-parallelize-loops = 2)	1.515	1.609	1.875	1.009	1	1.008
7	Clang (serial)	1.515	1.593	1.859	N.A	N.A	N.A
8	Clang (-fopenmp)	N.A	N.A	N.A	N.A	N.A	N.A

parallel cycles is half the serial cycles. In the case of do_across program each iteration depends on the previous one and there is no parallel potential and the speedup is reported as one. It should be noted that generating cycles after analysis is more reliable as compared to run times gathered by executing the programs, since it is prone to overheads of implementation and measurement. We have not provided the code for the programs used in the measurement due to limited space available.

7   Conclusion

This research project started out, with the objective of finding an ideal intermediate representation, for the domain of Program Transformation and Parallelization. We examined the published literature and prevailing solutions in the field. Not finding a suitable IR technology for our use, we designed ASIF, a three address representation as the intermediate representation for our Asterix compiler. ASIF IR is made up of two parts namely the *Executable Instructions* and *Annotations* and is designed to satisfy all the, Code Generation and Code Transformation and Parallel Conversion requirements. We presented arguments to show, how ASIF our three address instruction format, is a suitable intermediate representation, for the Program Transformation and Parallelization problem. We identified some popular transformations and parallel conversions that can benefit from the information emitted in the form of annotations, which complement the IR instructions generated with enough information for code generation. We also compared the technology from the competition such as LLVM, GIMPLE, WHIRL, MASM and NASM and contrasted them with ASIF, focusing on their fitness for use as IR for transformation and parallelization purposes. We also gathered speedup results for Asterix/ASIF and compared them with two other open source technologies. Based on the comparison and results, we conclude that ASIF is a superior technology that has been designed from the ground up, keeping the transformation and parallelization requirements in perspective, and comparable to the best-of-breed solutions offered by the competition and is better in many areas.

References

1. Hennessy, J.L., Patterson, D.A.: Computer Architecture, Fifth Edition: A Quantitative Approach, 5th edn. Morgan Kaufmann Publishers Inc., San Francisco (2011)
2. Faigin, K.A., Weatherford, S.A., Hoeflinger, J.P., Padua, D.A., Petersen, P.M.: The Polaris internal representation. Int. J. Parallel Prog. **22**(5), 553–586 (1994)
3. Giordano, M., Furnari, M.M.: HTGviz: a graphic tool for the synthesis of automatic and user-driven program parallelization in the compilation process. In: Polychronopoulos, C., Fukuda, K.J.A., Tomita, S. (eds.) ISHPC 1999. LNCS, vol. 1615, pp. 312–319. Springer, Heidelberg (1999). https://doi.org/10.1007/BFb0094932
4. Hall, M.W., et al.: Maximizing multiprocessor performance with the suif compiler. Computer **29**(12), 84–89 (1996)
5. SUIF. The suif 2 compiler system, February 2021. http://suif.stanford.edu/suif/suif2/
6. Belevantsev, A.A.: Multilevel static analysis for improving program quality. Program. Comput. Softw. **43**(6), 321–336 (2017)
7. Bastoul, C.: Code generation in the polyhedral model is easier than you think. In: Proceedings of the 13th International Conference on Parallel Architecture and Compilation Techniques, PACT 2004, pp. 7–16, September 2004
8. Trifunovic, K., Nuzman, D., Cohen, A., Zaks, A., Rosen, I.: Polyhedral-model guided loop-nest auto-vectorization. In: 18th International Conference on Parallel Architectures and Compilation Techniques, PACT 2009, pp. 327–337, September 2009

9. Pouchet, L., Bastoul, C., Cohen, A., Vasilache, N.: Iterative optimization in the polyhedral model: Part I, one-dimensional time. In: International Symposium on Code Generation and Optimization, CGO 2007, pp. 144–156, March 2007

10. Kandemir, M., Choudhary, A., Shenoy, N., Banerjee, P., Ramenujarn, J.: A linear algebra framework for automatic determination of optimal data layouts. IEEE Trans. Parallel Distrib. Syst. **10**(2), 115–135 (1999)

11. Fahringer, T., Scholz, B.: A unified symbolic evaluation framework for parallelizing compilers. IEEE Trans. Parallel Distrib. Syst. **11**(11), 1105–1125 (2000)

12. Saito, H., Stavrakos, N., Carroll, S., Polychronopoulos, C., Nicolau, A.: The design of the PROMIS compiler. In: Jähnichen, S. (ed.) CC 1999. LNCS, vol. 1575, pp. 214–228. Springer, Heidelberg (1999). https://doi.org/10.1007/978-3-540-49051-7_15

13. Suganuma, T., et al.: Overview of the IBM java just-in-time compiler. IBM Syst. J. **39**(1), 175–193 (2000)

14. Grosser, T., Groesslinger, A., Lengauer, C.: Polly–performing polyhedral optimizations on a low-level intermediate representation. Parallel Process. Lett. **22**(04), 1250010 (2012)

15. Zakai, A.: Emscripten: an LLVM-to-JavaScript compiler. In: Proceedings of the ACM International Conference Companion on Object Oriented Programming Systems Languages and Applications Companion, pp. 301–312 (2011)

16. Liao, C., Quinlan, D.J., Panas, T., de Supinski, B.R.: A ROSE-based OpenMP 3.0 research compiler supporting multiple runtime libraries. In: Sato, M., Hanawa, T., Müller, M.S., Chapman, B.M., de Supinski, B.R. (eds.) IWOMP 2010. LNCS, vol. 6132, pp. 15–28. Springer, Heidelberg (2010). https://doi.org/10.1007/978-3-642-13217-9_2

17. Dan, Q., Liao, C.: The rose source-to-source compiler infrastructure. In: Cetus Users and Compiler Infrastructure Workshop, in Conjunction with PACT, vol. 2011, p. 1. Citeseer (2011)

18. Lee, S.-I., Johnson, T.A., Eigenmann, R.: Cetus – an extensible compiler infrastructure for source-to-source transformation. In: Rauchwerger, L. (ed.) LCPC 2003. LNCS, vol. 2958, pp. 539–553. Springer, Heidelberg (2004). https://doi.org/10.1007/978-3-540-24644-2_35

19. Bae, H., et al.: The Cetus source-to-source compiler infrastructure: overview and evaluation. Int. J. Parallel Program. **41**(6), 753–767 (2013)

20. Gupta, S., Dutt, N., Gupta, R., Nicolau, A.: SPARK: a high-level synthesis framework for applying parallelizing compiler transformations. In: 2003 Proceedings of 16th International Conference on VLSI Design, pp. 461–466. IEEE (2003)

21. Lattner, C., Adve, V.: LLVM: a compilation framework for lifelong program analysis & transformation. In: Proceedings of the International Symposium on Code Generation and Optimization: Feedback-directed and Runtime Optimization, CGO 2004, pp. 75–86. IEEE Computer Society, Washington (2004)

22. LLVM.Developer. LLVM compiler infrastructure, February 2021. https://llvm.org/docs/

23. AMD.Developer. Open64 compiler developer guide, February 2021. https://developer.amd.com/compiler-developer-guide/

24. Gcc.gnu.org. GCC online documentation, February 2021. https://gcc.gnu.org/onlinedocs/

25. Intel.Corporation. Intel® C++ compiler classic developer guide and reference, February 2021. https://software.intel.com/content/www/us/en/develop/documentation/cpp-compiler-developer-guide-and-reference/top.html

26. Intel.Corporation. Intel® fortran compiler classic 2021.1 and intel®fortran compiler (beta) developer guide and reference, February 2021. https://software. intel.com/content/www/us/en/develop/documentation/fortran-compiler-oneapi-dev-guide-and-reference/top.html

27. Microsoft.Corporation. docs.microsoft.com, February 2021. https://docs.microsoft. com/en-us/

28. Microsoft.Corporation. Microsoft C++, C, and assembler documentation, February 2021. https://docs.microsoft.com/en-us/cpp/?view=msvc-160

29. PathScale. Inc., Ekopath documentation, February 2021. https://www.pathscale. com/documentation

30. PathScale. Inc., Ekopath user guide, February 2021. https://www.pathscale.com/ EKOPath-User-Guide

31. Kalyur, S., Nagaraja, G.S.: ParaCite: auto-parallelization of a sequential program using the program dependence graph. In: 2016 International Conference on Computation System and Information Technology for Sustainable Solutions (CSITSS), pp. 7–12, October 2016

32. Kalyur, S., Nagaraja, G.S.: Concerto: a program parallelization, orchestration and distribution infrastructure. In: 2017 2nd International Conference on Computational Systems and Information Technology for Sustainable Solution (CSITSS), pp. 204–209, December 2017

33. Kalyur, S., Nagaraja, G.S.: AIDE: an interactive environment for program transformation and parallelization. In: 2017 2nd International Conference on Computational Systems and Information Technology for Sustainable Solution (CSITSS), pp. 199–203, December 2017

34. Kalyur, S., Nagaraja, G.S.: Efficient graph algorithms for mapping tasks to processors. In: Haldorai, A., Ramu, A., Mohanram, S., Chen, M.-Y. (eds.) 2nd EAI International Conference on Big Data Innovation for Sustainable Cognitive Computing. EICC, pp. 467–491. Springer, Cham (2021). https://doi.org/10.1007/978-3-030-47560-4_35

35. Kalyur, S., Nagaraja, G.S.: Evaluation of graph algorithms for mapping tasks to processors. In: Haldorai, A., Ramu, A., Mohanram, S., Chen, M.-Y. (eds.) 2nd EAI International Conference on Big Data Innovation for Sustainable Cognitive Computing. EICC, pp. 423–448. Springer, Cham (2021). https://doi.org/10.1007/ 978-3-030-47560-4_33

36. Kalyur, S., Nagaraja, G.S.: CALIPER: a coarse grain parallel performance estimator and predictor. In: Miraz, M.H., Excell, P.S., Ware, A., Soomro, S., Ali, M. (eds.) iCETiC 2020. LNICST, vol. 332, pp. 16–39. Springer, Cham (2020). https:// doi.org/10.1007/978-3-030-60036-5_2

37. Kalyur, S., Nagaraja, G.S.: A survey of modeling techniques used in compiler design and implementation. In: 2016 International Conference on Computation System and Information Technology for Sustainable Solutions (CSITSS), pp. 355–358, October 2016

38. Kalyur, S., Nagaraja, G.S.: A taxonomy of methods and models used in program transformation and parallelization. In: Kumar, N., Venkatesha Prasad, R. (eds.) UBICNET 2019. LNICST, vol. 276, pp. 233–249. Springer, Cham (2019). https:// doi.org/10.1007/978-3-030-20615-4_18

39. Canedo, A., Sowa, M., Abderazek, B.A.: Quantitative evaluation of common subexpression elimination on queue machines. In: v Parallel Architectures, Algorithms, and Networks, I-SPAN 2008, pp. 25–30, May 2008

40. Rinker, R., Hammes, J., Najjar, W.A., Bohm, W., Draper, B.: Compiling image processing applications to reconfigurable hardware. In: 2000 Proceedings of IEEE International Conference on Application-Specific Systems, Architectures, and Processors, pp. 56–65 (2000)
41. Chen, L.-L., Wu, Y.: Aggressive compiler optimization and parallelization with thread-level speculation. In: 2003 Proceedings of 2003 International Conference on Parallel Processing, pp. 607–614, October 2003
42. Bohm, W., Najjar, W., Shankar, B., Roh, L.: An evaluation of coarse grain dataflow code generation strategies. In: 1993 Proceedings of Programming Models for Massively Parallel Computers, pp. 63–71, September 1993

Performance Comparison of VM Allocation and Selection Policies in an Integrated Fog-Cloud Environment

M. R. Shinu$^{(\boxtimes)}$ and M. Supriya

Department of Computer Science and Engineering, Amrita School of Engineering, Bengaluru, Amrita Vishwa Vidyapeetham, Bengaluru, India
{mr_shinu,m_supriya}@blr.amrita.edu

Abstract. Virtualization generates the virtual machine (VM) instances of software and hardware built on a single physical machine and helps to utilize the resources and infrastructure completely. VMs must be migrated to avoid over utilization and under utilization of hosts which can be done effectively by VM allocation and selection policies in resource management. Such requirements can be completely satisfied when using the fog enabled cloud systems. This paper proposes a combination of VM selection and allocation algorithms for integrated fog cloud system to improve the performance parameters like energy consumption, network use and cost. The results of the simulation show that the proposed strategies outperform the default VM selection and allocation algorithms in fog - cloud environment. The paper also presents the comparison of the above algorithms on the real-time data sets obtained from Google trace and PlanetLab.

Keywords: Cloud computing · Fog computing · VM allocation · VM selection

1 Introduction

As developments in the IT industry become more prominent, the future would focus on technologies that can offer accuracy and efficiency in real time [1]. Because of the fast growth and commercialization of information and communication networks, the percentage of cloud-connected devices is on the rise causing more bandwidth usage, cost and energy consumption [2]. This rise would also cause difficulty for cloud providers in offering uninterrupted services in turn leading to intermittent network connectivity issues especially for mobile wireless applications [3]. Due to the existence of large number of devices between the cloud data center and the end devices, there are machines aka fog nodes incorporated in between, to manage the storage and processing. The amount of data being produced and transferred between the cloud datacenter and the end users'

N. Kumar et al. (Eds.): UBICNET 2021, LNICST 383, pp. 169–184, 2021.
https://doi.org/10.1007/978-3-030-79276-3_13

increase day-by-day, causing high bandwidth consumption. This necessitates the need to minimize and optimize the bandwidth at the lower layer of networking architectures like fog nodes. Fog computing model is a decentralized paradigm that brings the source of interaction closer to computing nodes [4]. If the data being processed does not need high computation power, the processing can happen in the fog node while on the other case, the partial computation happens in the fog node and the partly processed data is sent to the cloud for further processing [5]. To be specific, the minimal processing and generating of responses can be done at the fog level and the data can be transmitted to the higher layers for further processing. Strict latency criteria for real-time services can also be addressed in fog effectively when compared to the cloud model. This interoperability feature of fog computing ensures massive support for various real-time applications of the smart world today [6]. Added advantages of fog computing include accessibility, reduction in bandwidth, improved response time, latency elimination, increased security, improved privacy etc. Fog computing claims to support the applications such as the internet of things (IoT), sixth generation (6G), artificial intelligence and machine learning [3].

With the tremendous increase and demand for various services on the cloud/fog computing platforms, the efficient use of resources has become a critical problem [7]. As the number of end devices is rising with each day, addressing the problems of resource management across networks is a major challenge and hence policies for resource selection and allocation play an important role [8]. The resource allocation policies have a significant effect on the performance of applications as well as the resource utilization [9]. Effective resource utilization can be achieved by virtualization which improves the availability of resources in turn leading to a better efficiency [10,11]. The technique of virtualization enables the sharing of resources through virtual machines among customers. Every virtual machine is isolated and is used to run applications, including its storage, processing capabilities and bandwidth for the specified requirements [12]. For maximum resource utilization, the optimal mapping of tasks to VMs and VMs to physical machines is important [13]. Efficient VM placement is aimed at improving efficiency, maximum utilization of resources and reducing energy usage in data centers without SLA violations [14]. Resource inefficiency and consumption of more bandwidth can occur if the VMs are not optimally placed in physical machines [9]. The allocation of resources is a complex process for cloud and fog services. One of the most key issues to be answered is to achieve an optimized selection and allocation of VMs to satisfy both users and providers [10]. This paper addresses the issues discussed and has the following major contributions

- Feasibility analysis of fog computing, VM allocation and selection policies
- Comparison of the cost, energy consumption and the network usage with the existing works
- Results implemented and compared with Google trace and PlanetLab dataset

The rest of the paper is arranged as follows: Sect. 2 analyses the related works. Section 3 outlines the description of the allocation and selection policies.

The algorithm and the experiment are discussed in Sect. 4. The performance evaluation is discussed in Sect. 5. The paper concludes with Sect. 6.

2 Related Works

Virtualization technology is an efficient way to optimize the use of resources and reduce data center energy costs due to which the virtual machine placement has created significant interest among researchers [15]. Many conventional methods are being used for cloud and fog environments for the VM placement. However, it suffers from poor performance, under utilization and over utilization of host, additional energy consumption, high network use and extra cost.

Chauhan et al. discusses the various algorithms for VM placement and VM selection, including Non-Heuristic and Heuristic-based algorithms, and analyzes their possible improvements that result in reduced SLA violations, fewer VM migrations and less energy consumption [16]. Ruan et al. suggest a new architecture for VM allocation and migration that allocates and migrates virtual machines based on host performance-to-power ratios in cloud by comparing three energy-efficient VM allocation and selection algorithms namely IqrMc (Inter quartile range/Maximum correlation), MadMmt (Median absolute deviation/Minimum migration time) and ThrRs (Static Threshold/ Random selection). The results show a 69.31% reduction in energy consumption for different types of host computers with less migration and shutdown times for cloud computing data centers, but experiences a slight degradation of performance [17].

Debashis et al. explains three different policies for virtual machine allocation, namely Serial VM Allocation Policy, Optimal VM Allocation Policy, and GS VM Allocation Policy, and shows the VM allocation procedure for the host. On comparing the VM allocation cost, the proposed work concludes that the Serial VM Allocation Policy outperforms the rest and is suitable for the real time placement [18]. Sandeep et al. implements India's smart grid piloted by Power grid Corporation of India with an intra-city network in 68 cities with a presence in 178 cities. Simulation is conducted to accommodate cloudlets from 178 cities/centers at most. The parameters namely VM image size, VM RAM, VM MIPS, VM bandwidth and cloudlet length have been examined on two VM provisioning policies- time-shared and space-shared. The findings indicate that the VM placement cost increases linearly for the considered parameters [19].

On comparing different types of cloud tasks and virtual machine allocation strategies, Xing et al. indicate that the greedy approach takes least time while the randomized allocation strategy takes maximum time for VM scheduling [20]. For VM allocation, Aneeba et al. suggests two heuristic reliant Vector Bin Packing algorithms called FFDmean and FFDmedian. The results thus obtained indicates that FFDmean and FFDmedian significantly outperformed two other existing algorithms, Dot-Product and L2 [21]. Mohammed Rashid et al. suggests numerous revamped VM placement algorithms and incorporated a strategy to migrate clustered VMs by considering both the usage of the CPU and the allotted RAM. An exhaustive Cloudsim based performance study conducted on VM positioning achieved remarkable improvements on the parameters considered and shows substantial improvement to the existing PABFDVM placement algorithm [22]. Variants of the above algorithms proposed by Anurag et al. suggests that Best Fit Decreasing with Minimum Migration Time (BFDMMT) combined with the MMT VM selection policy significantly outperforms other dynamic VM consolidation algorithms [23].

A new VM placement policy called Meets Performance (MP) proposed by Xiong et al. prefers to place a migratable VM on a host with the minimum correlation coefficient. Results when compared gives better scores for the parameters namely energy consumption, VM migration time and SLA violation percentage [24]. Beloglazov et al. compared the performances of various VM allocation policies against the methodology proposed by ZoltánÁdám et al. [25]. The findings show differences up to 66 % on the same real-world workload traces, thus emphasizing the importance of assessing the effectiveness of competing algorithms objectively.

Mehran et al. proposes an accurate micro-genetic VM allocation algorithm where the number of host shutdowns is far less than GA and heuristic PABFD algorithms. It has minimized power consumption and a reduction in violation of SLA [26]. Ming-Hua et al. suggests the VM allocation method that reduces the cost of communication between network providers under service quality conditions by considering the problem of virtual machine placement into a mixed-integer linear programming result of which improves computational performance [15]. Hatem et al. proposes Mixed Integer Linear Programming (MILP) based VM placement model to reduce the total energy consumption considering that variables such as workload of the VM, user profiles, and the proximity of fog nodes [27]. Qizhen Li et al. proposes two coordinated VM allocation methods based on SMDP to find an asymptotically optimal strategy of VM allocation to improve revenue for the service provider [28]. CloudBench, proposed by Mario et al., enhances the evaluation and implementation of VM allocation approaches in private clouds [29].

The literature review could be summarized as follows: various papers evaluated the performance for cloud systems, taking into a combination of account parameters such as cost, energy consumption, VM migration time, SLA violation, and duration of task completion. But many of the proposed papers fails to address the performance parameters namely network usage, energy consumption, and cost computation in fog nodes and hence is still an open research topic. The current implementations have not addressed the selection and allocation policies to meet expected service for efficient fog computing systems. This paper tries to address few of the issues and provides an environment for the feasibility of VM selection and allocation policies in fog environment. It also uses the proposed methodology and environment to compare the different VM allocation algorithms.

3   VM Placement

The main challenge for Virtual Machine allocation is to allocate VM requests from users to the physical machine by minimizing cost, energy consumption and network use without affecting the performance of the system. The effective allocation of virtual machines to physical machine is one of the optimization challenges to achieve complete utilization of resources in data centers [30]. In order to avoid a possible SLA violation, it must decide whether a host is overloaded and decide when to migrate VMs from the host. It must decide when a host is underutilized and when to move VMs from the host to avoid resource underutilization [25]. The VM allocation policies used in cloud computing are median absolute deviation (MAD), inter quartile range (IQR), local regression (LR), robust local regression (LRR) and Static Threshold (THR). There are four classifications of policy for VM selection which includes Maximum Correlation (MC), Migration Time Minimum (MMT), Minimum consumption (MU) and selection at random (RS). A VM that has the highest correlation is migrated by MC compared to the other VMs on the same host. The MMT migrates those VMs that take the least time to migrate. Without any rules, VMs generally migrate randomly in RS policy [31].

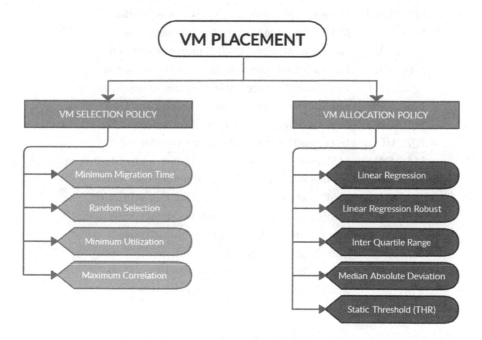

Fig. 1. VM placement

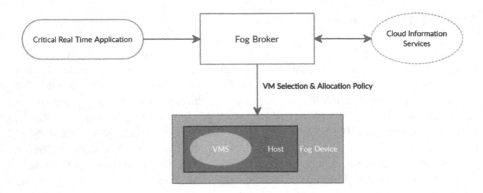

Fig. 2. Proposed model

Table 1. List of acronyms

LR/MMT	Linear regression/Minimum migration time
LR/RS	Linear regression/Random selection
LR/MU	Linear regression/Minimum utilization
LR/MC	Linear regression/Maximum correlation
LRR/MMT	Robust local regression/Minimum migration time
LRR/RS	Robust local regression/Random selection
LRR/MU	Robust local regression/Minimum utilization
LRR/MC	Robust local regression/Maximum correlation
IQR/MMT	Inter quartile range/Minimum migration time
IQR/RS	Inter quartile range/Random selection
IQR/MU	Inter quartile range/Minimum utilization
IQR/MC	Inter quartile range/Maximum correlation
MAD/MMT	Median absolute deviation/Minimum migration time
MAD/RS	Median absolute deviation/Random selection
MAD/MU	Median absolute deviation/Minimum utilization
MAD/MC	Median absolute deviation/Maximum correlation
THR/MMT	Static Threshold/Minimum migration time
THR/RS	Static Threshold/Random selection
THR/MU	Static Threshold/Minimum utilization
THR/MC	Static Threshold/Maximum correlation

4   System Model and Problem Formulation

This section describes the VM placement in the fog environment. Figure 1 depicts the VM selection and allocation policies considered for the proposed model. For the proposed model, the assumption is that every fog device has the data center characteristics such as host and VMs to provide processing capacity demanded by real-time applications as can be seen in Fig. 2. The proposed model implements the chosen VM selection and allocation policies and compares their performances on different workload traces which is described in the next section. To achieve results in the fog environment, all the combinations of VM selection and VM allocation policies have been considered. The VM selection and allocation policy combinations considered are LR/MMT, LR/RS, LR/MU, LR/MC, LRR/MMT, LRR/RS, LRR/MU, LRR/MC, IQR/MMT, IQR/RS, IQR/MU, IQR/MC, MAD/MMT, MAD/RS, MAD/MU, MAD/MC, THR/MMT, THR/RS, THR/MU, THR/MC, the expanded notations of which are listed in Table 1. The proposed model considers a single fog device with all the cloudlets corresponding to the user tasks.

5   Performance Evaluation

5.1   Experimental Setup

It would be extremely costly and difficult to carry out experiments on a real world environment. Repeatable tests in complex system conditions and user implementations to test the performance of placement policies are often difficult to conduct. iFogSim simulator comes to the rescue to model and simulate such large fog environments. The toolkit supports fog computing platforms and application provisioning environments for modelling and simulation and includes custom interfaces for the execution of VM allocation strategies and policies in inter-networked fog computing applications [32]. Hence, the proposed placement policies have been implemented and tested using the iFogSim toolkit. The proposed allocation and selection policies for integrated fog cloud system are compared with the resident default VM allocation algorithms.

5.2   Real Time Workloads - Google Trace and PlanetLab

This paper compares the different VM selection and allocation policies using the iFogSim simulator. To have a real time comparison, the paper uses the real-time workload traces obtained from open source and presents a comparison among the VM policies. The dataset for the working model is obtained from Google Cluster Workload Traces 2019 [33] and PlanetLab [34]. A short description of the dataset is presented below:

– Google Cluster Workload Traces 2019: This is a trace of the workloads for the month of May 2019 operating on eight Google Borg computing clusters. For the jobs running in those clusters, the trace describes every job submission, scheduling decision, and resource usage data. This has allowed a wide range of research for cluster schedulers and cloud computing to advance the state-of-the-art, and has been used to produce hundreds of analysis and studies. To incorporate this data set the Power package in iFogSim toolkit has been altered.

– PlanetLab data: This is a global network infrastructure implementation and evaluation platform. It makes use of distributed virtualization in which a slice of each service operates global resources of PlanetLab. On PlanetLab, multiple slices run concurrently, where slices serve as network-wide containers that isolate each others services. During 10 random days in March and April 2011, a series of CPU use traces from PlanetLab VMs were collected.

5.3　Performance Comparison

Different combinations of VM selection and allocation policies have been applied to the proposed model and the resultant outputs of the system have been compared and analyzed. Energy consumption, cost and network usage are the parameters considered for the evaluation of VM selection and allocation policy combinations described in Sect. 4. The Google trace and PlanetLab datasets are used for comparing various policies based on metrices such as energy consumption, network use and the cost as discussed below:

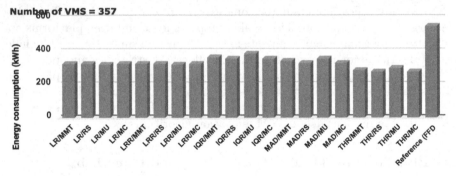

Fig. 3. Energy consumption comparison

Energy: Energy consumption is one of the critical elements in computing systems. This usually includes backup power supply infrastructure, environmental controls, security tools and communications. Improving energy efficiency has a substantial impact on system performance, availability, cost, reliability, and environmental protection. Evidently, energy optimization of the fog and cloud is crucial. Monitored energy consumption for all the VM selection and allocation policy combinations discussed in the preceding sections are compared with the

PlanetLab workload - Energy consumption

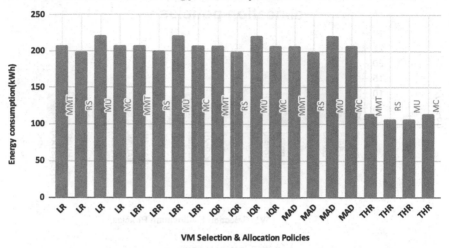

Fig. 4. Energy consumption comparison by applying PlanetLab workload

Google trace workload - Energy consumption

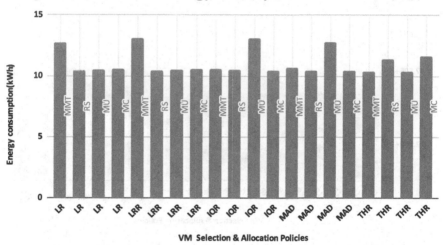

Fig. 5. Energy consumption comparison by applying Google trace

results presented in [35] and can be seen in Fig. 3. The energy consumption of
the VM selection and allocation policy combinations using PlanetLab workload
and Google trace workload is shown in Figs. 4 and 5 respectively.

Cost: Costs include hardware for networks, communications, and storage costs.
Savings on cost in fog computing is one of the primary challenges. Evaluated

Cost comparison of different VM selection and allocation policies

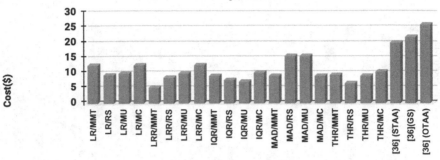

Fig. 6. Cost comparison

PlanetLab workload - Cost

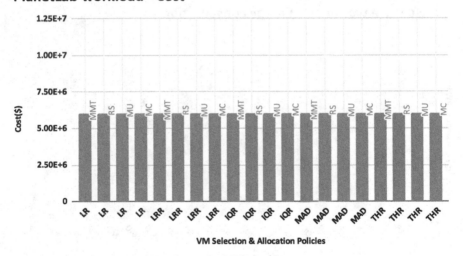

Fig. 7. Cost comparison by applying PlanetLab workload

costs for all the policy combinations are compared with the results presented in [36] and the same can be seen in Fig. 6. The cost of execution for the VM selection and allocation policy combinations using PlanetLab workload and Google trace workload is shown in Figs.7 and 8 respectively.

Network Use: If the edge of the network can handle the workload portion, relative to cloud operations, the latency can be dramatically improved. The

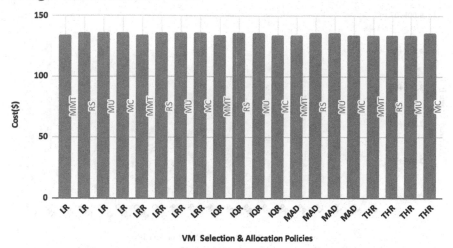

Fig. 8. Cost comparison by applying Google trace

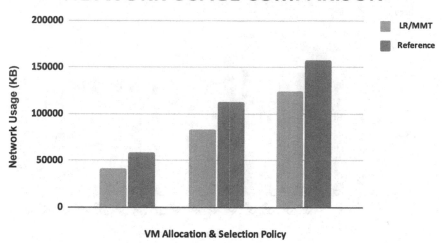

Fig. 9. Network use comparison

bandwidth of the edge-to-cloud is also preserved. The size of data transmission is substantially lowered by data pre-processing at the edge and fog systems. Many terminal devices are connected to the network and they deploy many database servers for service due to which bandwidth optimization is challenging. Figure 9 compares the results obtained by the proposed combinations against the results presented in [37]. Figures 10 and 11 present the real-time traces of the chosen dataset.

PlanetLab workload - Network Use

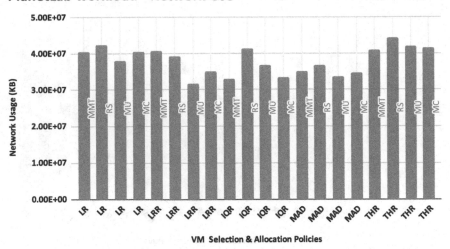

Fig. 10. Network use comparison applying PlanetLab workload

Google trace workload - Network Use

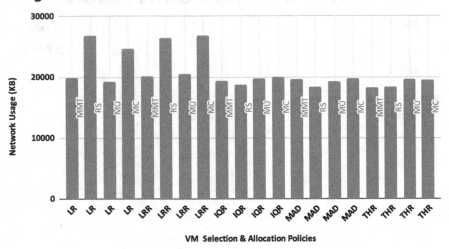

Fig. 11. Network use comparison by applying Google trace

On observing and analyzing the results thus obtained, it could be concluded that the fog model based implementation has a better performance in terms of the considered parameters than the standard cloud based implementation. Table 2 lists the best preferable VM allocation and selection policies for the real-time workload traces tested in this work.

Table 2. Comparison of VM allocation and selection policies

Workload	Network use	Energy consumption	Cost
Google trace	THR/MMT	THR/MU	THR (MU, RS), MAD (MC, MMT), IQR (MC, MMT), LRR/LR (MMT)
PlanetLab	LRR/MU	THR (RS, MU)	Similar cost of execution

The effectiveness of the proposed method for all the metrics considered in this work is calculated as $a = [(x - y)/y] * 100$ where x is the minimum parameter value obtained while applying VM allocation and selection policy combinations. y is the parameter value for the considered VM allocation policy from reference paper presented in the figures described above. a is the percentage of improvement for the metrics considered.

The findings show 58.8% decrease of energy use, 53.8% decrease of cost and 29.6% decrease of network usage in fog environment compared to existing algorithms, thus highlighting the significance of evaluating the performance of existing algorithms in fog environments.

6   Conclusion

The energy consumption, cost and network use of fog computing applications is an open research and needs to be addressed. This paper attempts to compare the various VM selection and allocation policies for the real-time workload traces by performing the computation on the fog nodes. Such a methodology when proved efficient, can be applied for many time-critical applications. In the simulated environment, while the policies presented have better efficiency, the application on the real-time must be studied. But on comparison, for energy parameter, THR/MU allocation yields a better performance when compared with all the combinations. While considering cost of execution, almost all allocation combinations shows similar results. For network use THR/MMT gives better results for Google trace and LRR/MU gives better results for PlanetLab. However, this is subjective and can vary from dataset to dataset.

References

1. Watters, A.: 10 emerging trends in information technology for 2020, February 2020. https://www.comptia.org/blog/10-emerging-trends-in-information-technology-for-2020
2. Sun, G., Zhou, R., Sun, J., Yu, H., Vasilakos, A.V.: Energy-efficient provisioning for service function chains to support delay-sensitive applications in network function virtualization. IEEE Internet Things J. **7**(7), 6116–6131 (2020)
3. Chiang, M., Zhang, T.: Fog and IoT: an overview of research opportunities. IEEE Internet Things J. **3**(6), 854–864 (2016)
4. Iyer, G.N., Veeravalli, B.: Cloud brokers. In: Encyclopedia of Cloud Computing, p. 372 (2016)
5. Prakash, P., Darshaun, K., Yaazhlene, P., Ganesh, M.V., Vasudha, B.: Fog computing: issues, challenges and future directions. Int. J. Electr. Comput. Engi. **7**(6), 3669 (2017)
6. Osanaiye, O., Chen, S., Yan, Z., Lu, R., Choo, K.-K.R., Dlodlo, M.: From cloud to fog computing: a review and a conceptual live VM migration framework. IEEE Access **5**, 8284–8300 (2017)
7. Gopinath, P.G., Vasudevan, S.K.: An in-depth analysis and study of load balancing techniques in the cloud computing environment. Proc. Comput. Sci. **50**, 427–432 (2015)
8. Zaidi, R.T.: Virtual machine allocation policy in cloud computing environment using cloudsim. Int. J. Electr. Comput. Eng. (2088–8708) **8**(1), 344–354 (2018)
9. Shabeera, T., Kumar, S.M., Salam, S.M., Krishnan, K.M.: Optimizing VM allocation and data placement for data-intensive applications in cloud using ACO metaheuristic algorithm. Int. J. Eng. Sci. Technol. **20**(2), 616–628 (2017)
10. Zhang, P., Zhou, M., Wang, X.: An intelligent optimization method for optimal virtual machine allocation in cloud data centers. IEEE Trans. Autom. Sci. Eng. **17**, 1725–1735 (2020)
11. Bourguiba, M., Haddadou, K., El Korbi, I., Pujolle, G.: Improving network I/O virtualization for cloud computing. IEEE Trans. Parallel Distrib. Syst. **25**(3), 673–681 (2013)
12. Deepika, T., Prakash, P.: Power consumption prediction in cloud data center using machine learning. Int. J. Electr. Comput. Eng. (IJECE) **10**(2), 1524–1532 (2020)
13. Bharathi, P.D., Prakash, P., Kiran, M.V.K: Energy efficient strategy for task allocation and vm placement in cloud environment. In: 2017 Innovations in Power and Advanced Computing Technologies (i-PACT), pp. 1–6. IEEE (2017)
14. Bharathi, P.D., Prakash, P., Kiran, M.V.K.: Virtual machine placement strategies in cloud computing. In: 2017 Innovations in Power and Advanced Computing Technologies (i-PACT) (2017)
15. Lin, M.-H., Tsai, J.-F., Hu, Y.-C., Su, T.-H.: Optimal allocation of virtual machines in cloud computing. Symmetry **10**(12), 756 (2018)
16. Chauhan, N., Rakesh, N., Matam, R.: Assessment on VM placement and VM selection strategies. In: Panigrahi, B.K., Hoda, M.N., Sharma, V., Goel, S. (eds.) Nature Inspired Computing. AISC, vol. 652, pp. 157–163. Springer, Singapore (2018). https://doi.org/10.1007/978-981-10-6747-1_18
17. Ruan, X., Chen, H., Tian, Y., Yin, S.: Virtual machine allocation and migration based on performance-to-power ratio in energy-efficient clouds. Futur. Gener. Comput. Syst. **100**, 380–394 (2019)

18. Das, D., Chanda, P.B., Biswas, S., Banerjee, S.: An approach towards analyzing various VM allocation policies in the domain of cloud computing. In: Mandal, J.K., Mukhopadhyay, S., Dutta, P., Dasgupta, K. (eds.) CICBA 2018. CCIS, vol. 1030, pp. 344–351. Springer, Singapore (2019). https://doi.org/10.1007/978-981-13-8578-0_27

19. Mehmi, S., Verma, H.K., Sangal, A.: Simulation modeling of cloud computing for smart grid using CloudSim. J. Electr. Syst. Inf. Technol. 4(1), 159–172 (2017)

20. Xu, X., Hu, H., Hu, N., Ying, W.: Cloud task and virtual machine allocation strategy in cloud computing environment. In: Lei, J., Wang, F.L., Li, M., Luo, Y. (eds.) NCIS 2012. CCIS, vol. 345, pp. 113–120. Springer, Heidelberg (2012). https://doi.org/10.1007/978-3-642-35211-9_15

21. Soomro, A.K., Shaikh, M.A., Kazi, H.: FFD variants for virtual machine placement in cloud computing data centers. Int. J. Adv. Comput. Sci. Appl. 8(10), 261–269 (2017)

22. Chowdhury, M.R., Mahmud, M.R., Rahman, R.M.: Clustered based VM placement strategies. In: 2015 IEEE/ACIS 14th International Conference on Computer and Information Science (ICIS), pp. 247–252. IEEE (2015)

23. Shrivastava, A., Patel, V., Rajak, S.: An energy efficient vm allocation using best fit decreasing minimum migration in cloud environment. Int. J. Eng. Sci. 4076, 4076–4082 (2017)

24. Fu, X., Zhou, C.: Virtual machine selection and placement for dynamic consolidation in cloud computing environment. Front. Comp. Sci. 9(2), 322–330 (2015)

25. Mann, Z.A., Szabó, M.: Which is the best algorithm for virtual machine placement optimization? Concurr. Comput.: Pract. Exp. 29(10), e4083 (2017)

26. Tarahomi, M., Izadi, M., Ghobaei-Arani, M.: An efficient power-aware VM allocation mechanism in cloud data centers: a micro genetic-based approach. Clust. Comput. 24(2), 1–16 (2020)

27. Alharbi, H.A., Elgorashi, T.E., Elmirghani, J.M.: Energy efficient virtual machines placement over cloud-fog network architecture. IEEE Access 8, 94 697–94 718 (2020)

28. Li, Q., Zhao, L., Gao, J., Liang, H., Zhao, L., Tang, X.: SMDP-based coordinated virtual machine allocations in cloud-fog computing systems. IEEE Internet Things J. 5(3), 1977–1988 (2018)

29. Gomez-Rodriguez, M.A., Sosa-Sosa, V.J., Carretero, J., Gonzalez, J.L.: Cloudbench: an integrated evaluation of VM placement algorithms in clouds. J. Supercomput. 76, 1–34 (2020)

30. Nwe, K.M., Zaw, Y.M.: Efficient mapping for VM allocation scheme in cloud data center. In: 2020 IEEE Conference on Computer Applications (ICCA), pp. 1–4 (2020)

31. Chowdhury, M.R., Mahmud, M.R., Rahman, R.M.: Implementation and performance analysis of various VM placement strategies in CloudSim. J. Cloud Comput. 4(1), 20 (2015)

32. Gupta, H., Vahid Dastjerdi, A., Ghosh, S. K., Buyya, R.: iFogSim: a toolkit for modeling and simulation of resource management techniques in the internet of things, edge and fog computing environments. Softw. Pract. Exp. 47(9), 12-75–1296 (2017)

33. Verma, A., Pedrosa, L., Korupolu, M.R., Oppenheimer, D., Tune, E., Wilkes, J.: Large-scale cluster management at Google with Borg. In: Proceedings of the European Conference on Computer Systems (EuroSys). Bordeaux, France (2015)

34. Peterson, L., Bavier, A., Fiuczynski, M.E., Muir, S.: Experiences building PlanetLab. In: Proceedings of the 7th Symposium on Operating Systems Design and Implementation, pp. 351–366 (2006)
35. Khalil, A., Arshad, M., Kazi, H.: FFD variants for virtual machine placement in cloud computing data centers. Int. J. Adv. Comput. Sci. Appl. 8(10), 261–269 (2017)
36. Das, D., Chanda, P.B., Biswas, S., Banerjee, S.: An approach towards analyzing various VM allocation policies in the domain of cloud computing. In: Mandal, J.K., Mukhopadhyay, S., Dutta, P., Dasgupta, K. (eds.) CICBA 2018. CCIS, vol. 1030, pp. 344–351. Springer, Singapore (2019). https://doi.org/10.1007/978-981-13-8578-0_27
37. Taneja, M., Davy, A.: Resource aware placement of IoT application modules in fog-cloud computing paradigm. In: 2017 IFIP/IEEE Symposium on Integrated Network and Service Management (IM) (2017)

Artificial Neural Network, Machine Learning and Emerging Applications

Fraud Detection in Credit Card Transaction Using ANN and SVM

Anchana Shaji$^{(\boxtimes)}$, Sumitra Binu, Akhil M. Nair, and Jossy George

Department of Computer Science, CHRIST (Deemed to be University), Bengaluru, India
anchana.shaji@science.christuniversity.in

Abstract. Digital Payment fraudulent cases have increased with the rapid growth of e-commerce. Masses use credit card payments for both online and day-to-day purchasing. Hence, payment fraud utilizes a billion-dollar business, and it is growing fast. The frauds use different patterns to make the transactions from the cardholder's account, making it difficult for the organization or the users to detect fraudulent transactions. The study's principal purpose is to develop an efficient supervised learning technique to detect credit card fraudulent transactions to minimize the customer's and organization's losses. The respective classification accuracy compares supervised learning techniques such as deep learning-based ANN and machine learning-based SVM models. This study's significant outcome is to find an efficient supervised learning technique with minimum computational time and maximum accuracy to identify the fraudulent act in credit card transactions to minimize the losses incurred by the consumers and banks.

Keywords: SMOTE · Artificial neural network · Support vector machine · Credit card fraud detection

1 Introduction

In the financial sector, frauds are becoming a significant issue harming the organization and individual user's socioeconomic status. The fraudsters exploit payment mode through cards and online transactions to steal cardholder's money using various techniques. The fraudsters can retrieve data from different websites about the details of cardholders. Due to this criminal activity, the detection of fraud is an essential need in society. Fraudsters find different techniques to achieve their goals. Various types of frauds such as CNP (Card Not Present), Skimming, Phishing, making duplicate cards, attacks by using the magnetic strips behind the card are the major fraud attacks in the network [1]. Attackers fetch the details of cardholders from different websites to steal the amount from cardholder's account. The available prevention techniques [2] for fraudulent transactions in credit cards were Manual Review, Negative and Positive lists, Card Verification Methods (CVM), and Payer Authentication. Due to changes in fraudster's techniques, the detection of fraudulent transactions is difficult in credit card transactions.

© ICST Institute for Computer Sciences, Social Informatics and Telecommunications Engineering 2021
Published by Springer Nature Switzerland AG 2021. All Rights Reserved
N. Kumar et al. (Eds.): UBICNET 2021, LNICST 383, pp. 187–197, 2021.
https://doi.org/10.1007/978-3-030-79276-3_14

There is an exponential increase in online fraudulent transactions in the recent past, utilizing various e-mail spoofing, phishing, cloning a card, etc. These fraudulent transactions are contributing to revenue loss for both the financial institutions and customers. Although researchers have proposed many techniques to spot fraudulent activities with various models, there are still many limitations in the existing models that need to be addressed. Hence, this issue must be addressed, and the losses are reduced by an efficient online credit card detection mechanism. The study aims to develop an efficient fraud detection model by comparing supervised learning techniques such as deep learning-based, Artificial Neural Network, and machine learning-based, Support Vector Machine. By calculating and comparing each model's computation time and classification accuracy, this study focuses on developing a model capable of detecting fraudulent transactions in various credit cards.

The study is organized as follows: Recent and Related works are observed in Sect. 2 that compares different classification models to detect fraudulent transactions. Section 3 introduces the proposed framework along with its validation. The evaluation of the proposed approach is demonstrated throughout Sect. 4. Section 5, focuses on the results of the study. Further Section, concludes the work and discusses future work.

2 Related Work

As credit card becomes the most popular payment mode both in online and offline shopping, fraud cases are also rising. Recent studies have focused on detecting these fraudulent transactions by various techniques. Thulasyammal Ramiah Pillai et al. [3] have implemented a Multi-layer perceptron deep learning model to detect fraud by various parameters. Among the parameters, the activation functions such as logistic function and hyperbolic tangent function provide high performance, and the proposed model attains about 82% sensitivity. Further research can be conducted by improving the model accuracy by new activation functions in balanced data with more advanced deep learning models. The optimum results could be observed while adding hidden layers and nodes to the network.

Debachudamani Prusti et al. [4] have compared the accuracy of different classification algorithms such as KNN, Extreme Learning Machine, Random Forest, Multilayer Perceptron, and Bagging Classifier. Moreover, the rate at which the models can identify the non-ethical transactions in credit cards was much faster enough. These individual classification models are hybridized in which the ensemble machine learning algorithms improve the model's performance in detecting fraud and non-fraud. Further studies can be done by implementing the model by utilizing real-time data, which leads to more efficient outputs. Chunzhi Wang et al. [5] proposed a Back Propagation (BP) neural network which detects frauds in networks. The authors propose a fraud detection algorithm using the Whale Optimization Algorithm to optimize the Neural Network algorithm with Back Propagation. This algorithm leads to improved accuracy, which contributes to the efficiency of the detection system. Saurabh C. Dubey et al. [6] have considered the real-time dataset for detecting fraud in credit card transactions. ANN with the Backpropagation technique is proposed with 99.96% of accuracy. The research can be extended further by integrating the proposed model with cloud services, which detects fraud faster using automated techniques.

Rimpal R. Popat et al. [7] have reviewed the credit card fraud detection by different methods such as Logistic Regression, deep learning, SVM, Naive Bayesian, Artificial Immune System, KNN, Decision Tree, and Genetic Algorithm. The work also discusses various types of fraud such as skimming, phishing, Card NOT Present (CNP), and Stolen Card. The patterns of fraud keep on changing, and hence it is not easy to detect them. Pradheepan Raghavan et al. [8] deal in evaluating Machine Learning techniques such as SVM, RF, KNN, and Deep Learning techniques such as Convolutional Neural Network, Deep Belief Networks (DBN), and Restricted Boltzmann Machine (RBM). The work has utilized German, Australian, and European datasets to evaluate each dataset type's suitable method. For large datasets, integrating SVM with CNN is the best method, and for small datasets, an ensemble model consisting of SVM, Random Forest, and KNN produces better results.

S. S. Harshini Padmanabhuni et al. [9] have implemented machine learning techniques such as SVM, KNN, Linear Regression, Adaboost, Decision Tree, Random Forest, classification with Neural Network, and PNN (Probabilistic Neural Network) were implemented. These techniques are compared with the generated hybridized models which include of SVM, KNN, Logistic Regression, Neural Network, and Decision Tree. Among these techniques, the hybridized model ensures accuracy of around 82.47%. Mrs.Vimala Devi. J et al. [10] have generated models like Random Forest, Decision Tree as well as SVM. They have compared an imbalanced dataset in detecting the fraudulent transactions of the credit card. Among these models, the Decision Tree has better accuracy. Further studies can be done after balancing the datasets by oversampling the dataset. Jasmine A Hudali et al. [11] have identified various types of credit card fraud and have reviewed different algorithms such as deep learning-based ANN and machine learning-based Decision Tree. While detecting such fraudulent transactions in credit cards, the card owner has the right to block the card, which will protect the corresponding user's privacy.Y. Sahin [12] has analyzed classification techniques such as SVM and Decision tree with real-time data. SVM technique with different kernel functions such as RBF, Polynomial, Sigmoid, and linear gives 99% accuracy in detecting the credit card fraud transactions that lead banks to reduce their risk.

Nana Kwame Gyamfi et al. [13] have analyzed several forms of fraud and the Support Vector Machine with Spark (SVM-S) technique has been implemented for the classification of customer's normal and fraudulent behaviour. This classification focuses more on the validity of new incoming transactions. Comparison of this technique with the Back-Propagation Network shows that the performance of SVM-S is more efficient than the Back-Propagation Network. Aihua Shen et al. [14] analyze and compares different classification problems such as Decision Tree, Logistic Regression, and Neural Network to provide an efficient model to detect fraudulent transactions. Evaluating the model ANN and Logistic Regression performance were the best models to spot fraudulent transactions to minimize the bank's risk.

3   Methodology

Supervised techniques learn the previous behaviour from the data and can detect real-time credit card frauds. A binary classification model which determines the fraudulent and non-fraudulent transactions in credit card has been chosen and considered. The labelled data, which includes fraudulent or non-fraudulent transactions, is passed into the model. The model is then trained with input-output pairs, including various independent variables and the corresponding target(output) variable. During the testing phase, the model predicts the label of the unseen test data. Hence, supervised learning techniques such as deep learning-based ANN and machine learning-based SVM are analyzed and compared.

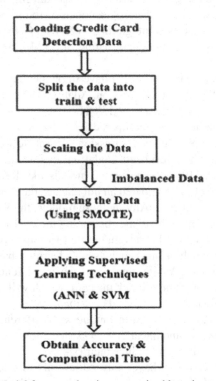

Fig. 1. Model framework using supervised learning techniques

Figure 1 shows the workflow of the proposed model. The imbalanced data is split into 70% for training and 30% for testing the data randomly. Hence, the data is scaled in the range of 0 and 1. Since the data is imbalanced, Synthetic Minority Oversampling Technique (SMOTE), an oversampling technique, is applied to balance the data. The SMOTE algorithm generates an arbitrary number of synthetic minority examples to balance the classifier's learning bias toward the minority class (i.e., fraudulent transactions) [15]. The resultant data is used to build supervised learning techniques such as deep learning-based Artificial Neural Network (ANN) and machine learning-based Support Vector Machine (SVM) models. Finally, these generated models would be evaluated by the classification's performance matrix, such as accuracy, precision, Recall, confusion matrix, F1-score, ROC-Curve, AUC-Curve, and computational time.

Overview of Artificial Neural Network (ANN)

Neural Network, as a classification technique, the training of the model is influenced by the initial parameter setting, bias, weight, and learning rate of the algorithm. This network's learning starts with some initial value, and the weights get renewed on each iteration. ANN works similar to the human brain. Propagated forward signals from the input layer to the hidden layer and then to the output layer would be implemented. The output layer's Job compares the output given by the hidden layer and the targeted output. If they are the same, that output will be updated. If an error is found, that will backpropagate the output, and then the weights get updated until the loss function is minimized [16].

This study' ANN framework consists of an input layer, four hidden layers, and an output layer. Two activation functions are compared with the corresponding accuracy of the model. In the first phase of the model training, ReLU (Rectified Linear Unit) Function will function as an activation function for the hidden layers. It acts like a linear activation function, which is easier to optimize [17]. The sigmoid function will be used as an activation function in the output layer for a binary classification problem to get the probability of the transaction in credit card is fraudulent or not. The second phase of the model training uses tanh (Hyperbolic Tangent) Function as an activation function for the hidden layers, and the sigmoid function will be used as an activation function in the output layer for a binary classification problem to get the probability of the credit card transaction is fraudulent or not. This model's significant parameters consist of the count of neurons in the input layer, hidden layer and output layer, activation function, batch size, learning rate, no. of epochs, dropout, and loss function as binary cross-entropy.

Overview of Support Vector Machine (SVM)

SVM considers all the input data and provides the output as a decision boundary as a Hyperplane that classifies both the fraudulent and non-fraudulent transactions. SVM learns a separating hyperplane, which maximizes the margin and produces good generalization ability. The distance between either nearest point is known as the margin. It finds a hyperplane that separates two classes (fraud or non-fraud) and has a high generalization capability that handles high-dimensional data. SVM transforms the attributes into high dimensional feature space and finds the optimal decision boundary that maximizes the classes' margin. A kernel function is used to remodel the dataset. This study uses a linear function that consists of only one hyperplane to classify the transactions into fraudulent or non-fraudulent.

Steps involved in developing an efficient supervised learning technique to detect credit card fraudulent transactions are as follows:

3.1 Dataset Used

The credit card data is available in the Kaggle data platform, in CSV format. It is used for research purposes that can be classified into fraudulent or non-fraudulent transactions. The dataset consists of 31 columns and 284,807 transactions of cardholders in Europe in 2013 for two days. Out of the 284,807 transactions, only 492 transactions are proportion fraudulent, making the data highly imbalanced. Two types of classes are identified in the dataset (i.e.) Fraudulent or non-fraudulent transactions.

Among the 31 features, 28 features named as V1, V2,......, V28 are the principal components derived from applying PCA, and the remaining three features, such as Time, Amount, and Class, have not been transformed by PCA. Class is a categorical variable describing whether the transaction is fraud or not. The amount is the transaction amount, and time denotes the interval elapsed between two successive transactions in the dataset. Because of the confidentiality of the data, original features are obfuscated [18]. However, we may guess that these features might be initially credit card number, type of transactions, transaction date-time, expiry date, purpose to use a credit card, CVV, credit card transaction history, transaction location, Job of the cardholders, cardholder name etc.

3.2 Data Splitting

The data is split into 70% for training and 30% for testing the new set of observations that predicts a fraudulent or non-fraudulent transaction randomly.

3.3 Data Scaling and Balancing

Except for the target variable, all other features are normalized into the range of 0 and 1 by using MinMax scaler from the sklearn library because it has only zeros and ones. The data has only 492 fraudulent transactions among 2,84,807 transactions, which shows that the data is highly imbalanced. Thus, the imbalanced data is balanced by the Synthetic Minority Oversampling Technique (SMOTE) which leads to an increase in the number of minority class observations (i.e., Fraudulent transactions) by creating synthetic observations [15]. Among the fraudulent transactions (minority class), k-Nearest Neighbors (KNNs) for each of the samples in the fraudulent transaction are identified. Then a line is drawn between the neighbors and generates random points on the line [19].

3.4 Feature Selection and Model Building

From the Oversampled data, the features are selected by feature selection methods by using mutual info classification function for the models. It is used to select the impacted features into the model, and the percentage of significant features is selected by using the select percentile function. In this study, supervised learning techniques such as deep learning-based ANN and machine learning-based SVM is implemented to identify the credit card fraudulent transactions with the training data.

3.5 Model Validation

In this study, the deep learning-based ANN model is validated through the validation_split argument in the Keras library's fit function which ranges from 0 to 1. It divides the training set accordingly by the variable' value. Hence, the first set is used for training, and the next set is used for validation after each epoch. For the machine learning-based SVM model, k-fold cross-validation is used for validating the model. It has a single parameter, k which subject to the count of different groups which splits the considered sample of data. The study considered the value of k as 10, i.e., 10-fold cross-validation which utilizes the unseen test data.

4   Evaluation Metrics

The generated algorithms can be evaluated by accuracy, Recall, precision, confusion matrix, and computational time to detect fraudulent credit card transactions.

The ANN model' first phase describes 97% of correctly predicted trained observations whether the transaction is fraud or not and 95% in testing the observations whether the transaction is fraud or not. Furthermore, the second phase of the ANN model describes 97% of correctly predicted trained observations whether the transaction is fraud or not and 98% in testing the observations on whether the transaction is a fraud or not. Also, SVM' training and testing accuracies are about 94% and 97% in predicting whether the observation is fraudulent or not.

Table 1 describes the Precision, Recall, and F1-Score of each model. Here Class 0 represents the non-fraudulent transactions, and Class 1 represents the fraudulent transactions. The first phase of the ANN model correctly predicts the fraudulent transactions with 3% and 100% for the non-fraudulent transactions. The second phase of the ANN model predicts fraudulent transactions with 8% and non-fraudulent transactions with 100% correctly. Also, the SVM model has 100% precision concerning ethical transactions and 6% for non-ethical transactions.

Moreover, for the first phase of the deep learning-based ANN model, the Recall is 93% correctly corresponding to fraudulent transactions, and the non-fraudulent transactions are about 95%. Moreover, for the second phase of the deep learning-based ANN model, the correctly concerning the fraudulent transactions are about 98%. Along with that, the correctly concerned non-fraudulent transactions are around 90%. For machine learning-based SVM, the Recall for fraudulent transactions is 92% and for non-fraudulent transactions is 98%.

The first phase of the deep learning-based ANN model's performance concerning the fraudulent transactions is 6% and of non-fraudulent transactions is about 97%. The second phase of the deep learning-based ANN model evaluates 15% for the fraudulent transactions and 99% corresponding to the non-fraudulent transactions. For machine learning-based SVM, the performance corresponding to the fraudulent transaction is 11% and to the non-fraudulent transactions is around 99%.

Table 1. Precision, Recall, and F1-Score of each model

Models		Precision	Recall	F1-Score
ANN_Model_1	Class 0	100%	95%	97%
	Class 1	30%	93%	60%
ANN_Model_2	Class 0	100%	98%	99%
	Class 1	80%	90%	15%
SVM	Class 0	100%	98%	99%
	Class 1	60%	92%	11%

The confusion matrix evaluates the generated classification algorithms' performance on a set of test data. Table 2. shows different ratios of the confusion matrix such as True Negative (TN) as the number of predicted non-fraudulent transactions, False Negative (FN) as the number of fraudulent transactions predicted as non-fraudulent transactions, False Positive (FP) as the number of non-fraudulent transactions predicted as fraudulent transactions and True Positive (TP) as the number of predicted fraudulent transactions.

Table 2. Ratios of Confusion Matrix of each model

Ratios	ANN_Model_1	ANN_Model_2	SVM
TN	81065	83943	83212
FN	9	13	11
FP	4242	1364	2095
TP	127	123	125

ROC-Curve is a probability curve obtained by plotting the True Positives on the y-axis and False Positives on the x-axis. AUC represents the measure of separability. Both phases of the Deep Learning-based ANN model have an AUC-score of 94%, and Machine Learning-based SVM model has an AUC-score of 95%. It implies that each model has a reasonable degree of separability as plotted in Fig. 2.

5 Results and Discussions

Each of the generated supervised learning techniques, such as deep learning-based ANN and machine learning-based SVM, is compared with their accuracy and computational time under the training and testing phase of each model is illustrated in Table 3.

Comparatively, it is identified that the second phase of the ANN model has maximum training and testing accuracy with minimum computational time in the training and testing phase of the model. According to the results, the preferable supervised technique for detecting fraudulent transactions in credit cards is Artificial Neural Network (ANN) with tanh (Hyperbolic Tangent) function as an activation function in the hidden layers.

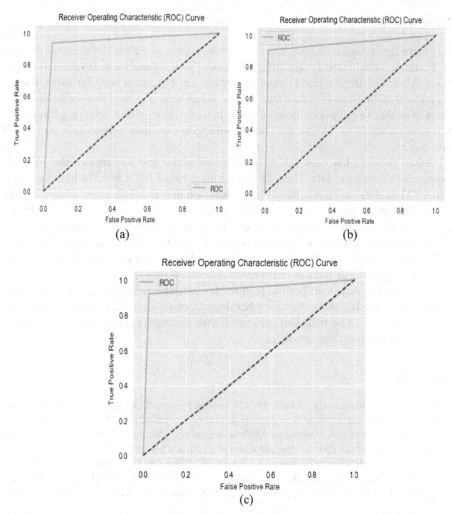

Fig. 2. (a) Roc-Auc Curve for ANN_ Model_1 (b) Roc-Auc Curve for ANN_ Model_2 (c) Roc-Auc Curve for SVM

Table 3. Comparison of each model with Accuaracy & Computational Time

Models	Accuracy		Computational time	
	Training accuracy	Testing accuracy	Training phase	Testing phase
ANN_Model_1	97%	95%	1865 s	0.10 s
ANN_Model_2	97%	98%	1816 s	0.04 s
SVM	94%	97%	2310 s	181.13 s

6 Conclusion

Credit Card fraud costs enormous losses to the customers as well as the financial companies. Fraudsters continuously find different ways of patterns and tactics to commit illegal actions. Thus, an efficient fraud detection system has become a need for users, banks, and financial institutions to reduce their losses. It has been observed that the imbalanced dataset available in the public domain gives biased results while detecting fraudulent credit card transactions. Hence, the considered credit card data is been split into 70:30 ratios.

Further, the data has been scaled by Minmax scaler and has applied the SMOTE technique to balance the data. Then, Deep Learning-based ANN and Machine Learning based SVM models are built to detect fraudulent and non-fraudulent transactions. ANN with the Backpropagation algorithm has the advantage of parallel processing capability. Also, SVM finds a hyperplane that separates two classes (fraud or non-fraud) and has a high generalization capability that handles high-dimensional data. The study shows that ANN with tanh as an activation function in hidden layers is an efficient classification to detect fraudulent transactions in credit cards.

Real-time data that describes the features for identifying the frauds in credit cards can be used for further study. Various other deep learning-based and machine learning-based techniques can be built and compared with different parameters, the activation function, loss function, and optimizer.

References

1. Saraswathi, E., Kulkarni, P., Khalil, M.N., Chandra Nigam, S.: Credit card fraud prediction and detection using artificial neural network and self-organizing maps. In: 2019 3rd International Conference on Computing Methodologies and Communication (ICCMC), Erode, India, pp. 1124–1128 (2019). https://doi.org/10.1109/ICCMC.2019.8819758
2. Roy, A., Sun, J., Mahoney, R., Alonzi, L., Adams, S., Beling, P.: Deep learning detecting fraud in credit card transactions. In: 2018 Systems and Information Engineering Design Symposium (SIEDS), Charlottesville, VA, pp. 129–134 (2018). https://doi.org/10.1109/SIEDS.2018.8374722
3. Pillai, T.R., Hashem, I.A.T., Brohi, S.N., Kaur, S., Marjani, M.: Credit card fraud detection using deep learning technique. In: 2018 Fourth International Conference on Advances in Computing, Communication & Automation (ICACCA), Subang Jaya, Malaysia, pp. 1–6 (2018). https://doi.org/10.1109/ICACCAF.2018.8776797
4. Prusti, D., Rath, S.K.: Fraudulent transaction detection in credit card by applying ensemble machine learning techniques. In: 2019 10th International Conference on Computing, Communication and Networking Technologies (ICCCNT), Kanpur, India, pp. 1–6 (2019). https://doi.org/10.1109/ICCCNT45670.2019.8944867
5. Wang, C., Wang, Y., Ye, Z., Yan, L., Cai, W., Pan, S.: Credit card fraud detection based on whale algorithm optimized BP neural network. In: 2018 13th International Conference on Computer Science & Education (ICCSE), Colombo, pp. 1–4 (2018). https://doi.org/10.1109/ICCSE.2018.8468855
6. Dubey, S.C., Mundhe, K.S., Kadam, A.A.: Credit card fraud detection using artificial neural network and backpropagation. In: 2020 4th International Conference on Intelligent Computing and Control Systems (ICICCS), Madurai, India, pp. 268–273 (2020). https://doi.org/10.1109/ICICCS48265.2020.9120957

7. Khine, A., Khin, H.W.: Credit card fraud detection using online boosting with extremely fast decision tree. In: 2020 IEEE Conference on Computer Applications (ICCA), Yangon, Myanmar, pp. 1–4 (2020). https://doi.org/10.1109/ICCA49400.2020.9022843

8. Raghavan, P., Gayar, N.E.: Fraud detection using machine learning and deep learning. In: 2019 International Conference on Computational Intelligence and Knowledge Economy (ICCIKE), Dubai, United Arab Emirates, pp. 334–339 (2019). https://doi.org/10.1109/ICCIKE47802.2019.9004231

9. Padmanabhuni, S.S.H., Kandukuri, A.S., Prusti, D., Rath, S.K.: Detecting default payment fraud in credit cards. In: 2019 IEEE International Conference on Intelligent Systems and Green Technology (ICISGT), Visakhapatnam, India, pp. 15–153 (2019). https://doi.org/10.1109/ICISGT44072.2019.00018

10. Vimala Devi, J., Kavitha, K.S.: Fraud detection in credit card transactions by using classification algorithms. In: 2017 International Conference on Current Trends in Computer, Electrical, Electronics and Communication (CTCEEC), Mysore, pp. 125–131 (2017). https://doi.org/10.1109/CTCEEC.2017.8455091

11. Kazemi, Z., Zarrabi, H.: Using deep networks for fraud detection in the credit card transactions. In: 2017 IEEE 4th International Conference on Knowledge-Based Engineering and Innovation (KBEI), Tehran, pp. 0630–0633 (2017). https://doi.org/10.1109/KBEI.2017.8324876

12. Popat, R.R., Chaudhary, J.: A survey on credit card fraud detection using machine learning. In: 2018 IEEE 2nd International Conference on Trends in Electronics and Informatics (ICOEI), Tirunelveli, pp. 1120–1125 (2018). https://doi.org/10.1109/ICOEI.2018.8553963

13. Gyamfi, N.K., Abdulai, J.: Bank fraud detection using support vector machine. In: 2018 IEEE 9th Annual Information Technology, Electronics and Mobile Communication Conference (IEMCON), Vancouver, BC, pp. 37–41 (2018). https://doi.org/10.1109/IEMCON.2018.8614994

14. Shen, R.T., Deng, Y.: Application of classification models on credit card fraud detection. In: 2007 International Conference on Service Systems and Service Management, Chengdu, pp. 1–4 (2007). https://doi.org/10.1109/ICSSSM.2007.4280163

15. He, H., Bai, Y., Garcia, E.A., Li, S.: ADASYN: adaptive synthetic sampling approach for imbalanced learning. In: 2008 IEEE International Joint Conference on Neural Networks (IEEE World Congress on Computational Fraud Detection in Credit Card Transaction Using Supervised Learning Techniques 60 Department of Data Science, Christ (Deemed to be University), Pune Lavasa Campus Intelligence), Hong Kong, pp. 1322–1328 (2008). https://doi.org/10.1109/IJCNN.2008.4633969

16. Brownlee, J.: A Gentle Introduction to the Rectified Linear Unit (ReLU), Machine Learning Mastery, 9 January 2019

17. Demla, N., Aggarwal, A.: Credit card fraud detection using SVM and reduction of false alarms. Int. J. Innov. Eng. Technol. (IJIET) 7(2), 176–182 (2016)

18. Kaggle, Credit Card Fraud Detection (2018)

19. Bhattacharyya, I.: SMOTE and ADASYN (Handling Imbalanced Data Set), Medium, 3 August 2018

Detection of Leukemia Using K-Means Clustering and Machine Learning

V. Lakshmi Thanmayi A$^{(\boxtimes)}$, Sunku Dharahas Reddy, and Sreeja Kochuvila

Department of Electronics and Communication Engineering, Amrita School of Engineering, Bengaluru, Amrita Vishwa Vidyapeetham, Bengaluru, India
avlt484@gmail.com, k_sreeja@blr.amrita.edu

Abstract. Leukemia or blood cancer is a common and serious disease across countries which is caused due to the sudden increase in White Blood Cells (WBCs) in blood. This increase in WBC is due to the production of immature or blast cells in the bone marrow of the affected person. Detection and diagnosis at early stage is important. Additionally, computer-aided diagnosis will enhance the process of detection with better accuracy. In this paper, we developed an algorithm for early-stage detection of leukemia using image processing. We also used machine learning classification techniques to classify between cancerous and non-cancerous cells. The algorithm uses K-means clustering for the segmentation of images and a linear Support Vector Machine (SVM) classifier for the classification. ALL-IDB data set has been used to validate the algorithm. A total of 368 images are used in the algorithm. Algorithm offers 95% of accuracy and an approximate 93% of precision.

Keywords: Leukemia · White Blood Cells · Machine learning · Support Vector Machine

1 Introduction

Cancer [1], the most common disease across the world, is caused due to the rapid multiplication of abnormal cells in the body. This reduces the functioning of the body. Cancer is of many types based on the cells attacked. Leukemia is a type of cancer which affects the white blood cells in the blood and thus called blood cancer [2]. Leukemia affects the bone marrow, the place of production of blood cells, causing an imbalance in the blood cell count. These increased abnormal cells lack the ability to fight against infection and affect the way healthy organs work.

Leukemia, if treated at the early stage, can stop the abnormal cells from increasing rapidly. Treatment at an early stage calls for the diagnosis of the disease at an early stage. The worldwide accepted methods in the diagnosis of leukemia include the physical examinations and lab tests [3]. During the physical examination, doctors look for swollen or bleeding gums and tiny rashes, which

© ICST Institute for Computer Sciences, Social Informatics and Telecommunications Engineering 2021
Published by Springer Nature Switzerland AG 2021. All Rights Reserved
N. Kumar et al. (Eds.): UBICNET 2021, LNICST 383, pp. 198–209, 2021.
https://doi.org/10.1007/978-3-030-79276-3_15

can be common with illnesses like flu. In the labs, the blood samples are analyzed by experts under powerful microscopes where the total blood count, including red blood cells, white blood cells, and platelets, are obtained. They also physically analyze the cells for the release of any substance that shows the presence of cancer. The diagnosis in the above methods can be effective only when the actual symptoms of the disease are seen. An algorithm for the detection of the disease even before any symptoms are visible is required. Many image processing algorithms have been developed to prove the presence of cancer at an early stage.

Digital image processing has a vast application in the diagnosis in the medical field. These image processing techniques have numerous advantages, for example, data flexibility, versatility, and estimation, information storing, and correspondence. Some significant machine learning classifiers include [4,5]: Linear classifiers, tree based classifiers, Support Vector Machine (SVM), Gaussian naive Bayes classifiers and Stochastic Gradient Descent (SGD) classifiers. SVM is a supervised learning strategy that can be utilized for both classification and regression. SVM accepts the input and output data points to create a hyperplane, also called a decision boundary which differentiates the input data based on the output classes. SVM can be used for both binary and multi-class classification.

In [6], a segmentation of nuclei cells and their classification algorithm is explained. Of the many segmentation techniques, the paper highlights the usage of K-means clustering method as a color based segmentation method, for image segmentation. The pixels are classified based on the *a and *b component of the L*a*b (luminosity, chromaticity layer-a, and chromaticity layer-b) color space. The paper features the significance of Hausdorff Dimensions during feature extraction. It helps in calculating roughness, perimeter and other parameters thus helping in the classification of images. Authors in [7] explains the classification of different types of cancer, Acute Myeloid Leukemia (AML) and Acute Lymphocytic Leukemia (ALL) are analysed depending on how leukemic cells look under the magnifying lens and the kind of cell included. While Chronic Myeloid Leukemia (CML) and Chronic Lymphocytic Leukemia (CLL) are analysed depending on the WBCs tally at the hour of conclusion. The presence of immature WBCs, or myoblasts in the blood and bone marrow is also used to conclude for AML and CML. This paper uses a mathematical operations based segmentation like addition and subtraction of various pre-processing of an image. The feature extraction of the segmented image is done using a MATLAB function "regionprops". Classification of the features is obtained using a SVM classifier for efficient results.

In this paper, we present an algorithm which gives an initial conclusion of the possible presence of cancer by a computerized analysis of digital, microscopic blood images. Image pre processing filters have been used to enhance the quality of the image for an output with lesser error. After the image is preprocessed, segmentation algorithms like: edge detection, watershed transform, K-means clustering, thresholding, are used to separate out the required part of the image from the whole image. We use the K-means clustering algorithm for the nuclei segmentation. Then, the features are extracted from the segmented images. These extracted features are used for classification in a machine learning algorithm. We show the binary classification of the data using a SVM classifier. The extracted features are considered as an input data and this input data is split into train and test sets which are used in classification of the images as cancerous and non-cancerous. The existing methods consider more number of features during the feature extraction stage for classification. In our proposed work we extract minimum number of features and analyse them to attain a greater accuracy. This reduces the computational power while extracting the features.

This paper is organized as follows. Section 2 gives few recent and relevant work in the area of leukemia detection. Section 3 describes the data sets and the steps to process the images. The proposed algorithm is explained in Sect. 4 followed by classification technique in Sect. 5. Results are analysed and discussed in Sect. 6. The conclusion and future scope is given in Sect. 7.

2   Few Recent and Relevant Work

In this section, we discuss few recent and very relevant work related to our work. A nucleus extraction method from the cytoplasm of the WBC as in [8] uses color conversion, intensity threshold and gradient method. An algorithm with color segmentation and Otsu's segmentation techniques for the separation process has been used. A computer aided diagnosis to recognize ALL kind of leukemia is created in [9]. In this, authors have discussed the proper feature extraction methods from the core of the WBCs for examination purposes. The paper deals with the appropriate feature extraction techniques from the nucleus of the WBCs for analysis purpose. Feature extraction is obtained using discrete cosine transform which helps in dividing the images into parts.

Authors in [10] used K-means clustering to extract the regions of interest along with basic enhancement, morphology filtering and segmentation technique. An adaptive histogram equalization is used for image pre-processing. The performance parameters in this paper have been analyzed using probability random index, this parameter gives us the exactness of segmentation. The paper explains in detail the various features in feature extraction. It explains the different set of values one gets during estimation which is helpful in the classification of the images. In [11] image pre-processing using median filters, conversion from Red Green Blue (RGB) to Hue, Saturation and Value (HSV) and thresholding have been performed. The boundaries of different image areas are found so that location, features and shape can be found in the image using the integral projection algorithm used in feature extraction. The paper uses watershed technique which is characterised by mountains and valleys. The mountains represent high intensity and valleys represent low intensity.

In [12], arithmetic operations and image enhancement techniques are used for segmentation of the nucleus from white blood cells. Also the problem of thresholding and K-means clustering can be solved by using this method. The image intensity is adjusted by using linear contrast stretching which segments out the nucleus. The blast cells from the normal lymphocyte cells are classified using a KNN classifier. In [13], the diagnosis of ALL type of cancer is explained by converting the RGB image into Cyan, Magenta, Yellow and Black (CMYK) scale as part of image pre-processing. Zack's algorithm, a triangle oriented threshold method, is used to segment out the WBCs. In this method a straight line is generated between the maximum and minimum value of the image histogram. After this an ideal threshold is determined and separation is completed utilizing the obtained threshold values.

In [14], important steps such as pre-processing, segmentation and match-making have been used. Otsu's image segmentation is used for segmenting the leukemia smear image database and matching of image pattern is done using Maximally Stable Extremal Regions (MSER). Pre-processing in this paper is completed using selective median filtering with unsharp masking and contrast enhancement techniques. MSER although having limited performance over blurred and/or textured images it has advantages over methods like moderate computational complexity, and it is more suitable for hardware implementation due to its algorithmic structure. Otsu's method maximises the between class variance by having an exhaustive search to evaluate this criteria. Segmentation in [15] is performed with morphological operators and Otsu's thresholding. Then, utilization of nucleus features with supervised KNN classifiers is used for the classification of the extracted features. In this process, the RGB images are converted into gray-scale to reduce the computational complications. Linear contrast enhancement and histogram equalization have been used for increasing the image intensity such that the total image, except for the nucleus, is brightened. Morphological erosion and closing operations to better the performance of segmentation and feature extraction processes is performed.

A fast correlation based filter has been used in [16] to select the most prominent gene for the feature extraction. This method is used to reduce the huge data set for classification. A SVM classifier is then implemented to classify the tumour cells. In [17] authors find different stages of CML using dynamic short distance pattern matching algorithm using the normal and abnormal gene sequences.

3   Typical Image Processing and Input Database

3.1   Typical Image Processing

The algorithm for the detection of leukemia normally involves multiple stages of image processing techniques which is shown in Fig. 1. These steps are used in sequence on the data set using MATLAB to obtain the best result. The conclusion of leukemia, either through the lab and manual tests or through image processing techniques, is obtained on microscopic blood smears. These blood smears are examined to check for any variations or abnormalities from normal blood cells. During blood tests a sample from the patient's blood is seen under a powerful magnifying lens or microscopes which amplify the blood smear as per the prerequisite. These microscopic blood samples are digitized into

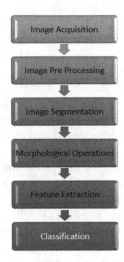

Fig. 1. Typical steps in image processing and classification

pictures with the goal that the image processing can be performed without any problem. These images of blood smears are accessible legitimately from the lab for a specific foundation purpose or are accessible online in picture banks where individuals can download for research purposes.

Image pre-processing is a way towards improving the image data, suppressing the undesired contortions, and enhancing the image features so that the analysis of the image at further stages of the algorithm yields a better error-free result. The images are made free of noise, de-blurred and other refinement processes are done as required by the procedure decided on an application use. Segmentation is a significant phase of the image processing process, since it removes the unwanted objects, separating the objects of our requirement, for additional processing. The major practical application of segmentation lies in the classification of the pixels where each pixel is assigned some labels. Pixels with similar labels share common charactcristics.

Morphological changes are some operations dependent on the picture shape. It is generally performed on binary images. It needs two information sources, the image on which morphological operations are performed and the structuring element or kernel which tells the type of operation to be performed. The two basic morphological operations are erosion and dilation. These two major operations have variations such as opening, closing, gradient and so on.

A picture has an immense informational collection that must be portrayed. The information in the picture is ordered into a lot of features which is called feature extraction. Feature extraction in machine learning, pattern recognition and in image processing, starts from initial data values obtained during the process and derives a new set of values which are more informative and non repetitive, helping in research, and classification which provide better human

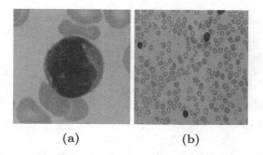

(a)   (b)

Fig. 2. Microscopic blood sample images (a) ALL-IDB1 (b) ALL-IDB2

conclusions. In this paper, the morphological features, related to the shape of the image, are analysed which helps in the classification in further steps.

After the features are extracted, the images or the data are classified into different classes or categories by analysing the features. An inter relation is established between these features through machine learning algorithms in order to classify the data easily and efficiently.

3.2 Input Database Images (IDB)

An online data set [18], ALL-IDB data set by Fabio Scotti from Università Degli Studi di Milano, has been used for leukemia detection in this paper. This data set includes ALL-IDB1 with 108 images. The nuclei from ALL-IDB1 have been separated in the form of images to form ALL-IDB2 which has 260 images. The algorithm has been implemented on both the data sets acquired. The example images of ALL-IDB1 and ALL-IDB2 are shown in Fig. 2.

4 Proposed Algorithm

This paper implements blind de-convolution and Gaussian noise removal methods for the refinement of the images. The acquired images are in an RGB space and are converted into a different color space such as L*a*b, CMYK, HSV for image segmentation. The RGB space is connected with the measure of light hitting the item and the genuine contrasts are not appropriately noticeable. This paper converts the image in RGB space to L*a*b space for the process of segmentation. The input and the converted image into L*a*b space is shown in the Fig. 3 and Fig. 4 respectively.

The images after pre-processing are segmented using K-means clustering algorithm to separate out the nucleus from the cytoplasm of the WBCs. K-means clustering is an iterative, color based segmentation method which is used to part a picture into K number of groups or clusters. K centroids are initialised first and every one of the pixels is mapped into its closest centroid esteem. In the wake of grouping all the pixels, a new centroid for each cluster is formed. This method uses Euclidean distance to calculate the distance between the pixels and

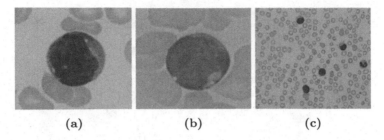

(a)   (b)   (c)

Fig. 3. Image samples used for pre-processing (a) Sample-001 (b) Sample-090 (c) Sample-022

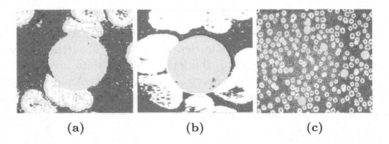

(a)   (b)   (c)

Fig. 4. Converted images from RGB to L*a*b (a) Sample-001 (b) Sample-090 (c) Sample-022

centroids to assign a pixel to a cluster with minimum distance. After the image is converted from RGB space, the colors are classified into K clusters. The process of clustering is carried out multiple times so as to avoid local minima. In this paper, the clustering process is performed three times where each component is separated out in each of the clustering processes. The nucleus component of the image is extracted after the clustering process is completed based on which cluster separates the nucleus best (as shown in Fig. 5).

After the process of segmentation, morphological operations are performed on the binary version of the segmented nucleus to make the image better analysable during feature extraction. Operations like closing, dilation, and opening of the image have been performed. Dilation is the process of adding pixels at the edges of the images in order to increase the white region of the images removing the shrunken look of the image. Opening is used to remove the minute noise, which is in the form of small white pixels over the black background, from the image. Closing is done when the small holes are there within the segmented image which are formed due to the variations in the color intensities. These holes are filled using the closing operation to form a solid, final image ready for feature extraction as shown in Fig. 6.

The segmented nucleus is analysed to extract the features. These features include energy, entropy, area, solidity and so on. In this paper we have analysed the nucleus to get the following features: area, perimeter, eccentricity, circularity,

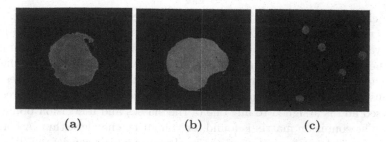

(a)   (b)   (c)

Fig. 5. Segmented blue nuclei (a) Sample-001 (b) Sample-090 (c) Sample-022 (Color figure online)

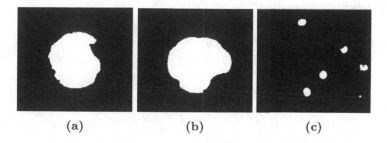

(a)   (b)   (c)

Fig. 6. Morphologically processed nuclei (a) Sample-001 (b) Sample-090 (c) Sample-022

and solidity. These features have been calculated for the images in ALL-IDB1 and ALL-IDB2 and the same were tabulated. Machine learning algorithms are implemented on these data values for the classification of the images.

5   Classification

A supervised machine learning model SVM has related calculations that is used for classification and regression. Given a set of input data, where each of the set belongs to anyone of the output classes, the SVM classifier trains a portion of input data to create a model which assigns the new values to one of the output classes. These points are represented in space such that they are divided by a hyper-plane also called decision boundary. The hyper plane creates a boundary between classes. Two classes as in this case, so that the data points lie on either side of the plane.

The kernel of the classifier that is used is defined by the user. Kernels represent the type of SVM classifier to be used. Kernels in SVM include linear, nonlinear, polynomial, radial basis function, and sigmoid. Gamma parameter is generally used in a non-linear SVM classifier where its value affects the Gaussian variance. A 'C' parameter is defined in an SVM classifier which regulates the error produced during training and testing. This is changed in a way to get minimum error in both the cases.

The classification of data in this paper has been done using a linear SVM classifier. The features extracted from the segmented images are in the form of integer values. The combination of each feature set representing the image as one of the two classes cancerous and non-cancerous. These values form the input data for the classifier which are trained based on the output class values where cancerous is considered as 1 and non cancerous is considered as 0. The test data is analysed and is assigned to any one of the classes and a decision boundary is formed. The confusion matrix is found for validating the algorithm. A confusion matrix is a tabular representation of the classifier which consists of True Positive (TF), True Negative (TN), False Positive (FP), False Negative (FN). The accuracy, precision, specificity and recall is calculated from the confusion matrix using the following formulae.

$$Accuracy = \frac{TP + TN}{Total}$$

$$Precision = \frac{TP}{TP + FP}$$

$$Specificity = \frac{TN}{TN + FP}$$

$$Recall = \frac{TP}{TP + FN}$$

6   Results and Discussion

The image processing steps are carried out on an ALL-IDB data set as explained in Fig. 1. The ALL-IDB1 and ALL-IDB2 images as in Fig. 3 have been pre processed to generate images as shown in Fig. 4. These images have been segmented to separate out the nucleus using K-means clustering process Fig. 5. The segmented nucleus is morphologically processed to improve the pixel rate and the final image in Fig. 6 is analysed for feature extraction using MATLAB. A linear SVM classifier has been implemented in order to classify the images to compute accuracy and validate the algorithm.

180 images from ALL-IDB1 and 260 images from ALL-IDB2 have been analysed. 80% of these have been used for training and 20% for testing purposes. Figure 7(a) and Fig. 7(b) represent the confusion matrix for ALL-IDB1 and ALL-IDB2 respectively. The accuracy, precision, specificity and recall were calculated from the confusion matrix as shown in Table 1. Accuracy determines how often the classifier is correct. Precision and recall refer to the percentage of the results which are relevant and the percentage of total relevant results correctly classified by the algorithm respectively.

Accuracy gives the overall performance of the classifier. Hence different classifiers were used to calculate the accuracy and the results are compared in Table 2. A highest accuracy of 95.45% is obtained for ALL-IDB1 and 95% accuracy for ALL-IDB2 using the proposed algorithm.

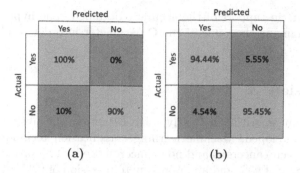

Fig. 7. Confusion matrices (a) ALL-IDB1 (b) ALL-IDB2

Table 1. Summary of different parameters.

Metrics	ALL-IDB1	ALL-IDB2
Accuracy	95.45%	95%
Precision	92.30%	94.44%
Specificity	90%	95.45%
Recall	100%	94.44%

Table 2. Comparison of accuracy using different classifiers

Classifier	Accuracy for ALL-IDB1	Accuracy for ALL-IDB2
SVM	95.45%	95%
KNN	81.8%	87.5%
Decision tree	90.69%	92.5%
Ada boost	90.9%	93.3%

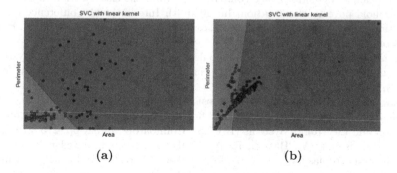

Fig. 8. Decision boundaries for Support Vector Classifier (SVC) (a) ALL-IDB1 (b) ALL-IDB2

The data points of the two classes are visualised in Fig. 8b where a hyper plane or decision boundary is drawn in order to separate the two classes. This hyper plane acts in support of the accuracy obtained. The hyper plane can be adjusted by changing the value of the C parameter in order to obtain a better accuracy.

7   Conclusion

In this work, an algorithm for early-stage detection of leukemia using image processing is developed. Machine learning classification techniques were used to classify between cancerous and non-cancerous cells. The proposed algorithm offers an accuracy of 95% and an approximate precision of 93% after performing image processing operations on a set of 368 blood images.

The availability of proper data set of images is necessary for better results. Getting such a uniform data set is sometimes challenging. Choosing a right method in each of the steps is essential to get accurate results. Though a lot of research is going on in this field, doctors and other biologists are reluctant on accepting and supporting this method due a chance of wrong result. This research can be extended further in order to increase the accuracy such that it can be considered as a diagnostic method for early detection of cancer.

References

1. WHO report on cancer: setting priorities, investing wisely and providing care for all. World Health Organization, Geneva (2020). Licence: CC BY-NC-SA 3.0 IGO
2. Canadian Cancer Society: Leukemia, understanding your diagnosis
3. Narayanan, U., Unnikrishnan, A., Paul, V., Joseph, S.: A survey on various supervised classification algorithms. In: 2017 International Conference on Energy, Communication, Data Analytics and Soft Computing (ICECDS), Chennai, pp. 2118–2124 (2017). https://doi.org/10.1109/ICECDS.2017.8389824
4. Talingdan, J.A.: Performance comparison of different classification algorithms for household poverty classification. In: 2019 4th International Conference on Information Systems Engineering (ICISE), Shanghai, China, pp. 11–15 (2019). https://doi.org/10.1109/ICISE.2019.00010
5. Agaian, S., Madhukar, M., Chronopoulos, A.T.: Automated screening system for acute myelogenous leukemia detection in blood microscopic images. IEEE Syst. J. 8(3), 995–1004 (2014)
6. Dharani, T., Hariprasath, S.: Diagnosis of leukemia and its types using digital image processing techniques. In: 2018 3rd International Conference on Communication and Electronics Systems (ICCES), Coimbatore, India, pp. 275–279 (2018)
7. Tran, V., Ismail, W., Hassan, R., Yoshitaka, A.: An automated method for the nuclei and cytoplasm of Acute Myeloid Leukemia detection in blood smear images. In: 2016 World Automation Congress (WAC), Rio Grande, pp. 1–6 (2016)
8. Mishra, S., Sharma, L., Majhi, B., Sa, P.K.: Microscopic image classification using DCT for the detection of acute lymphoblastic leukemia (ALL). In: Raman, B., Kumar, S., Roy, P.P., Sen, D. (eds.) Proceedings of International Conference on Computer Vision and Image Processing. AISC, vol. 459, pp. 171–180. Springer, Singapore (2017). https://doi.org/10.1007/978-981-10-2104-6_16

9. Kumar, S., Mishra, S., Asthana, P., Pragya: Automated detection of acute leukemia using K-mean clustering algorithm. In: Bhatia, S., Mishra, K., Tiwari, S., Singh, V. (eds.) Advances in Computer and Computational Sciences. Advances in Intelligent Systems and Computing, vol. 554, pp. 655–670. Springer, Singapore (2018). https://doi.org/10.1007/978-981-10-3773-3_64

10. Sigit, R., Bachtiar, M.M., Fikri, M.I.: Identification of leukemia diseases based on microscopic human blood cells using image processing. In: 2018 International Conference on Applied Engineering (ICAE), Batam, pp. 1–5 (2018)

11. Choudhary, R.R., Sharma, S., Meena, G.: Detection of leukemia in human blood samples through image processing. In: Bhattacharyya, P., Sastry, H., Marriboyina, V., Sharma, R. (eds.) NGCT 2017. Communications in Computer and Information Science, vol. 828, pp. 824–834. Springer, Singapore (2018). https://doi.org/ 10.1007/978-981-10-8660-1_61

12. Shafique, S., Tehsin, S., Anas, S., Masud, F.: Computer-assisted acute lymphoblastic leukemia detection and diagnosis. In: 2019 2nd International Conference on Communication, Computing and Digital systems (C-CODE), Islamabad, Pakistan, pp. 184–189 (2019)

13. Rege, M.V., Abdulkareem, M.B., Gaikwad, S., Gawli, B.W.: Automatic leukemia identification system using otsu image segmentation and mser approach for microscopic smear image database. In: 2018 Second International Conference on Inventive Communication and Computational Technologies (ICICCT), Coimbatore, pp. 267–272 (2018)

14. Umamaheswari, D., Geetha, S.: Segmentation and classification of acute lymphoblastic leukemia cells tooled with digital image processing and ML techniques. In: 2018 Second International Conference on Intelligent Computing and Control Systems (ICICCS), Madurai, India, pp. 1336–1341 (2018)

15. Bhagya, T., Anand, K., Kanchana, D.S., Remya, A.A.S.: Analysis of image segmentation algorithms for the effective detection of leukemic cells. In: 2019 3rd International Conference on Trends in Electronics and Informatics (ICOEI), Tirunelveli, India, pp. 1232–1236 (2019). https://doi.org/10.1109/ICOEI.2019.8862696

16. Kavitha, K.R., Gopinath, A., Gopi, M.: Applying improved SVM classifier for leukemia cancer classification using FCBF. In: 2017 International Conference on Advances in Computing, Communications and Informatics (ICACCI), Udupi, pp. 61–66 (2017)

17. Ananya, B., Prabisha, A., Kanjana, V.: Novel approach to find the various stages of chronic myeloid leukemia using dynamic short distance pattern matching algorithm. In: 2018 3rd International Conference for Convergence in Technology (I2CT), Pune, pp. 1–5 (2018)

18. Donida Labati, R., Piuri, V., Scotti, F.: All-IDB website. University of Milan, Departement of Information Technologies. http://www.dti.unimi.it/fscotti/all

An Analysis and Implementation of a Deep Learning Model for Image Steganography

Raksha Ramakotti$^{(\boxtimes)}$ and Surekha Paneerselvam

Department of Electrical and Electronics Engineering, Amrita School of Engineering, Bengaluru, Amrita Vishwa Vidyapeetham, Bengaluru, India
bl.en.p2ebs19009@bl.students.amrita.edu, p_surekha@blr.amrita.edu

Abstract. Steganography is the technique that involves hiding a secret data in an appropriate carrier. The major challenge involved in steganography is to ensure that the hidden data does not attract any attention towards it and hence works under the assumption that if the secret feature is visible, then the point of attack is evident. In this work, a novel deep learning model is designed to perform digital image steganography. The dataset used to train the model is Common Object in Context (COCO). An analysis is conducted based on batch size hyper-parameter, to evaluate the performance of the model. Also, the effect of using grayscale and color images on the evaluation metrics of the model is estimated. The analysis was orchestrated by evaluating the average Peak Signal to Noise Ratio (PSNR) and Structural Similarity Index (SSIM) of the trained images. The analysis has produced state-of-the-art results with optimized parametric values and has boosted computational efficiency producing a promising architecture to perform steganography.

Keywords: Image steganography · Convolutional neural networks · COCO dataset · Batch size · Peak signal to noise ratio · Structural similarity index

1 Introduction

Steganography is the art of covered or hidden writing where the goal is to covertly communicate a digital message. The word steganography is derived from two Greek words, steganós, meaning "covered or concealed", and graphia, meaning "writing" [1]. The main goal here is to communicate securely in a completely undetectable manner and to avoid drawing suspicion to the transmission of a hidden data. Hence, the nature of the information format and its quantity plays an important role. The data can be hidden in basic formats like: audio, video, text and images [2]. However, images are considered to be standard carriers because a text message cannot hide bulky data, audio is sensitive to noise and video steganography involves extensive pre-processing and as a data format per se, it is too heavy to analyse [3].

© ICST Institute for Computer Sciences, Social Informatics and Telecommunications Engineering 2021
Published by Springer Nature Switzerland AG 2021. All Rights Reserved
N. Kumar et al. (Eds.): UBICNET 2021, LNICST 383, pp. 210–224, 2021.
https://doi.org/10.1007/978-3-030-79276-3_16

The revolution in digital information has proved the need to send the message in a safe manner. Primarily, three such techniques have come into existence, i.e., cryptography, watermarking and steganography [4,5]. Though cryptography and watermarking are widely used, they showcase a few limitations. Regardless how strong is the cryptographic- encryption algorithm, it provides a scope of being decoded. In case of watermarking, the capacity of information chosen is limited by the application. It is steganography that potentially and effectively bridges these gaps. The advantage of steganography when compared to other methods is that the trace of secret information is unknown [6]. Media files are ideal for steganographic transmission because of their large size. The change is so subtle that someone who is not specifically looking for it is unlikely to notice the alteration.

In the previous work many image processing techniques have been used to implement the concept of steganography. Broadly, these methods can be categorised into three namely spatial domain, transform domain and machine learning techniques. Spatial domain aims to represent a grayscale image as a 2-D matrix or a color image as a 3-D vector of 2-D matrices. Least significant bit (LSB) steganography is a primary and fundamental method that works by replacing the LSBs of selected pixels in the cover image with secret message bits. Other popular spatial domain techniques include Edge Based Data Embedding and Random Pixel Embedding Methods [7,8]. The major transform domain techniques imparted for steganography include Discrete Cosine Transform (DCT) [9], Discrete Fourier Transform (DFT) and Discrete Wavelet Transform (DWT). Setiadi et al. have proposed an algorithm that conglomerates DCT alongside with OTP encryption [10]. Another novel ensemble technique that combines DCT and DWT are proposed in [11,12]. To implement image steganography, the DFT coefficients are modulated such that the secret information can be safely stored in a cover image. At the encoder end, the modulation presents the steganographic image. At the decoder side, this image is decomposed into frequency elements. In [13], Mandal proposes a DFT based image steganographic algorithm with a capacity of embedding payload of 0.75 bpB.

Machine Learning algorithms are mathematical models based on a set of training data which are used in the decision making process without a distinct instruction. Literature survey suggests that these models are broadly used in classification and regression [14,15]. Deep learning is a subset of Machine Learning that mimics the workings of the human brain in processing data for use in detecting objects, recognizing speech, and making decisions, to name a few [16–19]. The edge presented by Deep Learning is that it evinces the ability to learn without human supervision, drawing from data that is both unstructured and unlabelled. Convolutional Neural Network (CNN) is a special type of neural network that is specially designed to implicitly understand the intrinsic properties of images. With input being images, these networks are trained to perform a particular functionality to obtain the desired output. The handshake of neural networks to perform steganography has proven to improve the performance, robustness and efficiency of secret image communication. In [20], Baluja presents a robust model to perform color-color secret-cover image steganography trained using ImageNet dataset and evaluated his model obtaining Squared Summed Error values of cover and secret images using StegExpose Tool. The authors in [21,22], portray neural network models, emphasising on the fact that using grayscale images reduce the payload of the secret image on the cover image while hiding the data. In [23], the authors present a comparative study on LSB substitution technique and CNN based architecture. Wu et al.'s whole processing pipeline consists of two almost identical neural network structures responsible for encoding and decoding.

In this work, a deep learning model has been analysed and implemented to perform image steganography using CNN. The model is trained using COCO dataset and is analysed using Peak Signal to Noise Ratio (PSNR) and Structural Similarity Index (SSIM). Section 2 elucidates the methodology and presents a brief description on CNN and the dataset. Furthermore, Sect. 3 unfurls the discussion of analysis and results. The paper culminates with Sect. 4 which talks about the conclusion and future scope of this work.

2   Methodology

In this section, a brief description on CNN and the dataset used is provided. Along with this, proposed architecture for steganography is also presented.

2.1   Convolutional Neural Networks

Convolutional Neural Networks (CNNs) [20,21,23] are deep learning models that take a series of input images, assign importance (learnable weights and biases) to various aspects/objects in the image and be able to differentiate one image from the other. Unlike the primitive image methods, the filters and characteristics are not hand-engineered, instead the model is trained to learn and adapt these features to perform the desired functionality. The architecture portrayed by CNN is analogous to that of the connectivity design of neurons in the human brain. CNNs are designed to capture the spatial and temporal dependencies in an image through the application of relevant filters. These models are trained to understand the complexity and sophistication involved in images.

While designing an end to end encoder-decoder CNN model that performs steganography, the following properties are to be noted [24]:

- High Capacity: Amount of information can be embedded into an image.
- Perceptual Transparency: The property of a steganographic image that ensures the quality of the statistical properties of secret and cover image.
- Robustness: After embedding, data should stay intact if stego-image goes into some transformation such as cropping, scaling, filtering and addition of noise.
- Computation Complexity: The measure of computational expensiveness for designing a steganographic model.
- Temper Resistance: The degree with which the secret image is prone to being externally attacked.

2.2 Generalized Architecture for Steganography

The general architecture for steganography using CNN is presented in Fig. 1. Each of the blocks are further explained in the following subsections.

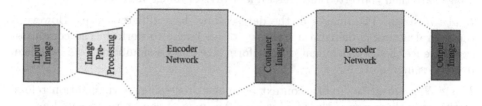

Fig. 1. Block diagram of a general architecture for steganography

Input Image. The input to the steganographic model is a raw image dataset. These images are digital in nature as they can be expressed in terms of a finite set of digital values. The images can be RGB (color) images or Grayscale in nature. Each image is described in (m × n × p) format, where (m × n) represents the size of the image and p represents the number of color channels. Therefore, it takes the value of 1 in case of grayscale image and 3 for a color image.

The smallest element of an image is called a pixel, which in case of color models take the value between 0–255. Thus, the color channels can be represented as an array of 8 bits. This shows that the image can be represented as a function,

$$b = f(m, n) \qquad (1)$$

where the value b represents the pixel coordinate value at that particular point.

Dataset Description. COCO stands for Common Object in Context. It is simply a conglomeration of everyday objects captured from everyday scenes. This adds some "context" to the objects captured in the scenes. COCO provides multi-object labeling, segmentation mask annotations, image captioning, key-point detection and panoptic segmentation annotations with a total of 81 categories, making it a very versatile, flexible and multi-purpose dataset. This dataset being open-source was particularly chosen as it introduces the concept of generalization amongst images through non-iconic images [25]. In this work, a subset of 2000 images were used to conduct the analysis.

Image Preprocessing. Image preprocessing is one of the most fundamental and important processes in any image related problem statement. Unlike the conventional digital image processing techniques, where preprocessing of the image is done through a set of algorithms and mathematical calculations, the construction of CNNs removes these computational and mathematical complexities involved. The outlook of preprocessing presented here is in the context of steganography.

Image Reading. The dataset is identified from its storage location path. The raw images are then converted into referable arrays of image matrices.

Image Resizing. The images in the dataset may have different sizes. However, the neural networks demand a base size of an image to perform the computations. Hence the establishment of a uniform size plays a significant role in image preprocessing.

Image Normalization. In the context of image processing, normalization refers to changing the range of the pixel values. An image I can be represented as

$$I = \sum_{i=0}^{M}\sum_{j=0}^{N}P(i,j) \tag{2}$$

here M and N are equal to 255 and P represents the pixel value at the ith and j^{th} coordinate in the image surface.

Since the range is too big, having this representation for the pixel size may increase mathematical complexity, computational time, increase the storage space and reduce the working efficiency of the model. However, while designing the models, the data scientist must ensure that the cost is not too expensive. Hence image normalization is performed on the same image such that the features of the image and its constructional sophistication is not lost. Thus, Eq. 2 can be reframed as

$$I_{norm} = \sum_{i=0}^{1}\sum_{j=0}^{1}P(i,j) \tag{3}$$

Dataset Splitting As a part of preprocessing technique that is specifically pertaining to training a steganographic model, the images are split into secret and carrier images. In some of the test cases, the secret images are converted to grayscale in this step.

Steganographic Model Description. The end-to-end process of steganography as showcased in Fig. 2 can be split into two:

Encoder. This process has two important purposes each of which is performed by two neural networks:

- Composition Network: This network takes the secret image as its input image. The purpose of having a dedicated network for a secret image is to understand the composition of the image. This network is trained to extract the features of the secret image and present it to the Conceal Network in such a way that there is slight to nil trace of the secret image on the carrier image. The output of the Composition Network is a transformed secret image which is fed to the Conceal Network.

- Conceal Network: It is in this network where the actual functionality of steganography is performed. The input to the Conceal Network consists of two things, one the transformed secret image and the other is the cover image. The model is to be trained in such a way that the statistical property of the cover and or the secret image is not distorted in the process. Since images provide a high capacity to hide another message, the trade-off between information heaviness and transparency must be taken care of ensuring that the very purpose of steganography is fulfilled. The output of this network is a container image. The model should be accurately trained in such a way that the container image looks similar to that of the carrier image.

Decoder. The process of revealing the secret image from the container image is known as steganalysis. This is done by the decoder which consists of the Exhibit Network. The presence of this network is what makes the complete process of steganography to be comprehensive and complete. The task of the Exhibit Network is to remove the carrier image without distorting the structural integrity of the secret image. Strikingly, the container image is the input to the Exhibit Network and the output image is to be as similar as the secret image.

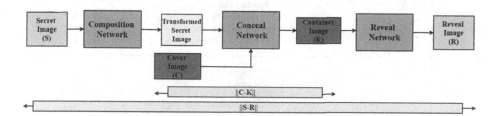

Fig. 2. End to end steganographic model architecture

Proposed Neural Network Architecture. The model architecture as described in Fig. 2 includes Composition, Conceal and Exhibit Networks. As shown in the figure, these networks are concatenated and trained together. The model's structure is primarily designed based on defining certain variables called hyper-parameters which are tabulated in Table 1.

The complete detailed model is presented in Fig. 3 Though the architecture borrows the structure from Baluja's model [20], there are some distinct differences. One, the kernel size of uniform (3×3) is utilized instead of (3×3), (4×4) and (5×5) because:

- Using even sized kernels does not produce a computational simplicity as all the previous layer pixels would not be symmetrically around the output pixel. This would lead to distortions as the center pixel cannot be interpolated.

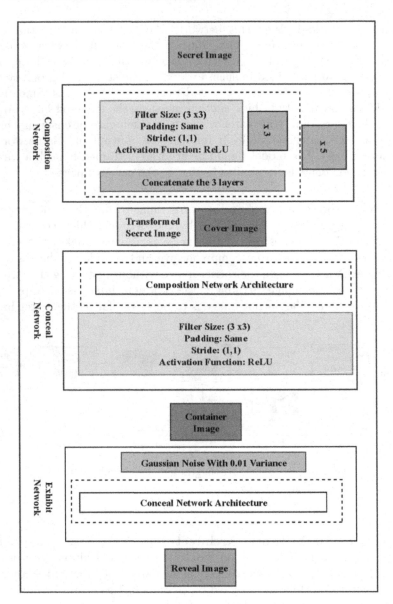

Fig. 3. Detailed depiction of proposed model architecture

- Using large kernel sizes such as (5 × 5) would increase the number of weights to be back-propagated which would lead to long training time and using more computational resources to do so.

Table 1. Hyperparameter tabulation for the proposed work

Hyper-parameter	Value
Number of layers	60 (20 in each network)
Strides	(1, 1)
Padding	Same
Kernel size	(3 × 3) in all layers
Activation function	ReLU
Epochs	100
Batch size	10–50
Learning rate	0.001

The kernel in a CNN model is a weighted matrix that is used to extract certain features of the image. This kernel is slid across the input image and convolution operation is performed such that the desired operation for which the network is designed is successfully performed.

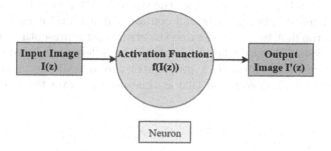

Fig. 4. Depiction of the working of activation function

The number of kernels in each layer is defined by the number of neurons. Each unit consists of a fixed number of neurons. In Baluja's model, it is described that each layer in each unit consists of 50 neurons. However, in the scope of proposing a low-parameter neural network, the architecture of the model here is constructed differently. The layers of the first unit have 50 neurons each. With each passing unit 10 neurons are reduced in order to afford the computational cost. The layers of the last unit thereby consist of 10 neurons each.

As shown in Fig. 4, the filtered image is passed to a mathematical "gateway" known as the activation function. It aids in transforming the input into a desired output through the mathematical definition of the function. The activation function used is ReLU which stands for Rectified Linear Unit. As shown in Fig. 5 it is a nonlinear piece-wise function, where the negative inputs are ceiled to 0 and for 0 and other positive inputs obey the mathematical formula,

$$b = f(0, n) \qquad (4)$$

where n belongs to positive integers.

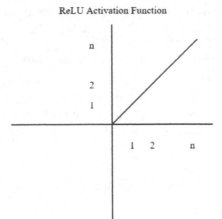

Fig. 5. Graph of ReLU Activation Function

As depicted in Fig. 2 error term i.e., the absolute difference of C and K, showcases the difference between cover and container image and it back propagates only through the first two networks and the error term representing the absolute difference of S and R,is calculated by taking the difference between secret and the reveal image. This term back propagates through all the three networks. Thereby, the entire loss equation that is thus used to train the model is given by,

$$L(C, K, S, R) = ||C - K|| + ||S - R|| \tag{5}$$

3 Analysis and Results

The evaluation metrics used to estimate the performance of the model analysed model architecture is presented in this section. COCO dataset was analyzed for cases where the secret images were color and grayscale. All the models were implemented in python 3 script in Google Colab. Tensorflow framework is used to model CNNs.

3.1 Evaluation Metrics

In Image Processing and Computer Vision applications, where both the input and the output are digital images, the quantifying evaluation metrics used to determine the performance of the are PSNR and SSIM. The details are provided in the following subsections.

Peak Signal to Noise Ratio. The Peak Signal to Noise Ratio (PSNR) [26] represents the ratio between the maximum power value of a signal and the power of distorting noise that affects the quality of its representation. The higher the PSNR, the better the quality of the reconstructed image. It is measured in decibels. The formula for PSNR is given as,

$$PSNR = 10log_{10}(\frac{MAX_{y_{ij}}^2}{MSE}) \tag{6}$$

where, MSE stands for Mean Squared Error and the numerator term represents the maximum possible pixel value of the image. When the pixels are represented using 8 bits per sample, this is 255.

Structural Similarity Index. The Structural Similarity Index (SSIM) [27] is an enduring metric that quantifies the degradation of an image quality due to extensive image processing or modelling. SSIM is given by,

$$SSIM(I,O) = (\frac{(2m_I m_O + c_1)(2s_{IO} + c_2)}{(m_I^2 + m_O^2 + c_1)(s_I^2 + s_O^2 + c_2)}) \tag{7}$$

where,
· m_I m represents the average of the input image I
· m_O represents the average of the output image O
· s_I represents the variance of the input image O
· s_O represents the variance of the output image O
· s_{IO} represents the covariance of the images I and O

It is to be noted that the evaluation metrics were calculated separately between the Cover and Container Images at the encoder end and Secret and Reveal Images at the decoder end. In each of these cases, Cover and Secret Images will be treated as the input image I and the corresponding Container and Reveal Images as the output image O, which is represented in the Eq. 7.

3.2 Performance Analysis of the Steganographic Model

Image Dataset: Cover and Secret Images Are Color. The model is constructed with the ReLU activation function in all the layers. The batch sizes are varied between 10 and 50. Figure 6. shows the variation of Encoder and Decoder Networks' PSNR with respect to batch size for COCO dataset, where both the cover and secret images are color images.

It can be observed that the encoder model showed an almost consistent performance with respect to PSNR, with its values in the range 26.91 dB and 27.18 dB. The decoder however showcased a slight descending drop in the PSNR value, when the batch size was increased. At batch size = 30, both the encoder and the decoder network an approximate value of 27 dB. The average PSNR value showcased by both the networks were quite similar too, i.e., 27.02 dB and 26.98 dB. However, at batch size =10, both encoder and decoder value gave the best result,

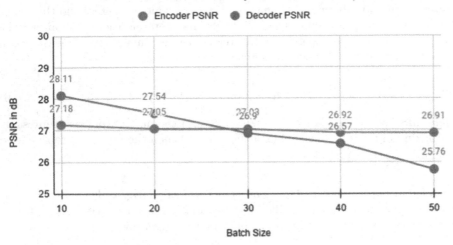

Fig. 6. Performance analysis of COCO dataset PSNR with respect to batch size

Table 2. Evaluation metrics tabulation of COCO dataset when cover and secret images are color

Batch size	Encoder PSNR (dB)	Decoder PSNR (dB)	Encoder SSIM	Decoder SSIM
10	27.18	28.11	0.94	0.93
20	27.05	27.54	0.94	0.93
30	27.03	26.90	0.94	0.93
40	26.92	26.57	0.94	0.93
50	26.91	25.76	0.94	0.93

i.e., 27.18 dB and 28.11 dB respectively. Since the COCO dataset introduced the characteristic of generalization, the stability in the PSNR values could be visibly observed from the graph. Interestingly, this consistency can be observed with the SSIM parameter as well. For all the batch sizes, the model reflected an SSIM of 0.94 and 0.93 for the container image and reconstructed image respectively. These values are tabulated in Table 2.

Image Dataset: Cover Images Are Color and Secret Images Are Grayscale. The same architecture showcased in Fig. 3 was trained for a scenario where the secret image was grayscale. The values of the evaluation metrics are tabulated in Table 3.

The trade-off between the encoder and the decoder performance can evidently be observed here. As portrayed in Fig. 7, the encoder PSNR drops when the secret image is grayscale. This is because the encoder network tends to treat the transformed grayscale image as a seemingly noise component, rather than the signal itself. The COCO dataset heavily suffers on this criteria. The average performance of the PSNR encoder for all the batch sizes was calculated to be

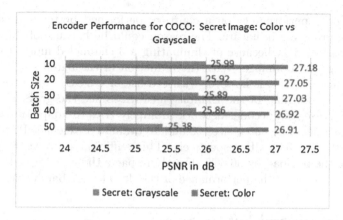

Fig. 7. Performance analysis of encoder with respect to batch size for color vs grayscale secret images for COCO dataset

Table 3. Evaluation metrics tabulation of COCO dataset when cover images are color and secret images are grayscale

Batch size	Encoder PSNR (dB)	Decoder PSNR (dB)	Encoder SSIM	Decoder SSIM
10	25.99	34.6	0.93	0.98
20	25.92	33.06	0.93	0.98
30	25.89	29.4	0.93	0.98
40	25.86	28.82	0.93	0.98
50	25.38	28.54	0.93	0.97

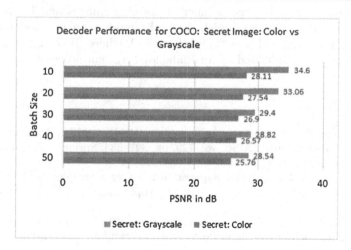

Fig. 8. Performance analysis of decoder with respect to batch size for color vs grayscale Secret images for COCO dataset

25.81 dB, which has dropped by 4.48% when compared to that of the average performance of PSNR encoders when color images are used as secret images.

The decoder performance on a contrary performed better with the secret image being grayscale in nature. The reason could be because of the reduction in the payload by, i.e., because of eliminating a 3 channeled image to expose a 1 channeled grayscale image. With the reduction in the batch size, the performance was improved. It is however with the use of COCO dataset, it could be understood the working of the Exhibit Network. Since the grayscale image is treated as a noise, the decoder network promptly tries to adapt the working of a denoising model. The batch sizes 20 and 10 evaluated excellent PSNR results, i.e., 33.06 dB and 34.6 dB respectively. This sufficiently proves the effect of reduction of the payload by 66.66% directly impacts the working of the decoder model. Figure 8 shows the performance of the decoder for batch sizes 10 to 50.

4   Conclusion and Future Scope

The implementation of a novel model for steganography has been analyzed successfully. The proposed model was designed and trained with the COCO dataset, which was used to induce the property of generalization. It was observed that for a batch size of 10, the models performed the best. COCO was consistent with its PSNR and SSIM values i.e., around 27 dB and 0.935 respectively. When the secret images were converted to grayscale, there was a dip in the encoder performance as the model treated the transformed secret images as noise components. On the contrary, with the reduction in the payload on the decoder end, "denoising" functionality proved to be an efficient task. The current model can be improved by increasing computational efficiency, training the model with more iterations, increasing the number of training images and further tuning the hyper-parameters of the model. Also, the model can be condensed to fit into the needs of being called a 'TinyML' model, where the complex process of steganography is implemented by an Embedded System processor.

References

1. Pund-Dange, S.: Steganography: a survey. In: Bokhari, M.U., Agrawal, N., Saini, D. (eds.) Cyber Security. AISC, vol. 729, pp. 327–333. Springer, Singapore (2018). https://doi.org/10.1007/978-981-10-8536-9_32
2. Febryan, A., Purboyo, T.W., Saputra, R.E.: Steganography methods on text, audio, image and video: a survey. Int. J. Appl. Eng. Res. **12**(21), 10485–10490 (2017)
3. Menon, N.: A survey on image steganography. In: 2017 International Conference on Technological Advancements in Power and Energy (TAP Energy), pp. 1–5. IEEE, December 2017
4. Cheddad, A., Condell, J., Curran, K., Mc Kevitt, P.: Digital image steganography: survey and analysis of current methods. Signal Process. **90**(3), 727–752 (2010)
5. Al-Yousuf, F.Q.A., Din, R.: Review on secured data capabilities of cryptography, steganography, and watermarking domain. Indonesian J. Electr. Eng. Comput. Sci. (IJEECS) **17**(2), 1053–1059 (2020)

6. Pillai, B., Mounika, M., Rao, P.J., Sriram, P.: Image steganography method using k-means clustering and encryption techniques. In: 2016 International Conference on Advances in Computing, Communications and Informatics (ICACCI), pp. 1206–1211. IEEE, September 2016

7. Hussain, M., Wahab, A.W.A., Idris, Y.I.B., Ho, A.T., Jung, K.H.: Image steganography in spatial domain: a survey. Signal Process.: Image Commun. **65**, 46–66 (2018)

8. Wang, Z.H., Chang, C.C., Li, M.C.: Optimizing least-significant-bit substitution using cat swarm optimization strategy. Inf. Sci. **192**, 98–108 (2012)

9. Kadhim, I.J., Premaratne, P., Vial, P.J., Halloran, B.: Comprehensive survey of image steganography: techniques, evaluations, and trends in future research. Neurocomputing **335**, 299–326 (2019)

10. Rachmawanto, E.H., Sari, C.A.: Secure image steganography algorithm based on DCT with OTP encryption. J. Appl. Intell. Syst. **2**(1), 1–11 (2017)

11. Sari, W.S., Rachmawanto, E.H., Sari, C.A.: A good performance OTP encryption image based on DCT-DWT steganography. Telkomnika **15**(4), 1987–1995 (2017)

12. Kumar, V., Kumar, D.: Digital image steganography based on combination of DCT and DWT. In: Das, V.V., Vijaykumar, R. (eds.) ICT 2010. CCIS, vol. 101, pp. 596–601. Springer, Heidelberg (2010). https://doi.org/10.1007/978-3-642-15766-0_102

13. Mandal, J.K., Khamrui, A.: A genetic algorithm based steganography in frequency domain (GASFD). In: 2011 International Conference on Communication and Industrial Application, Kolkata, India, pp. 1–4 (2011)

14. Raksha, R., Surekha, P.: A cohesive farm monitoring and wild animal warning prototype system using IoT and machine learning. In: 2020 International Conference on Smart Technologies in Computing, Electrical and Electronics (ICSTCEE), pp. 472–476. IEEE, October 2020

15. Se, S., Vinayakumar, R., Kumar, M.A., Soman, K.P.: Predicting the sentimental reviews in Tamil movie using machine learning algorithms. Indian J. Sci. Technol. **9**(45), 1–5 (2016)

16. Zhao, Z.Q., Zheng, P., Xu, S.T., Wu, X.: Object detection with deep learning: a review. IEEE Trans. Neural Netw. Learn. Syst. **30**(11), 3212–3232 (2019)

17. Zhang, Z., Geiger, J., Pohjalainen, J., Mousa, A.E.D., Jin, W., Schuller, B.: Deep learning for environmentally robust speech recognition: an overview of recent developments. ACM Trans. Intell. Syst. Technol (TIST). **9**(5), 1–28 (2018)

18. Ganesan, S., Krichen, M., Alroobaea, R., KP, S.: Robust Malware Detection using Residual Attention Network (2020)

19. Poorna, S.S., et al.: Computer vision aided study for melanoma detection: a deep learning versus conventional supervised learning approach. In: Pati, B., Panigrahi, C.R., Buyya, R., Li, K.-C. (eds.) Advanced Computing and Intelligent Engineering. AISC, vol. 1082, pp. 75–83. Springer, Singapore (2020). https://doi.org/10.1007/978-981-15-1081-6_7

20. Baluja, S.: Hiding images in plain sight: deep steganography. In: Advances in Neural Information Processing Systems, pp. 2069–2079 2017)

21. Rahim, R., Nadeem, S.: End-to-end trained CNN encoder-decoder networks for image steganography. In: Proceedings of the European Conference on Computer Vision (ECCV) (2018)

22. Zhang, R., Dong, S., Liu, J.: Invisible steganography via generative adversarial networks. Multimed. Tools Appl **78**(7), 8559–8575 (2018). https://doi.org/10.1007/s11042-018-6951-z

23. Wu, P., Yang, Y., Li, X.: Stegnet: mega image steganography capacity with deep convolutional network. Future Internet **10**(6), 54 (2018)

24. Pandikumar, T., Gebreslassie, T.: Information security using image based steganography. Int. Res. J. Eng. Technol. (IRJET) **3**(06), 2839–2844 (2016)
25. Lin, T.-Y., et al.: Microsoft COCO: common objects in context. In: Fleet, D., Pajdla, T., Schiele, B., Tuytelaars, T. (eds.) ECCV 2014. LNCS, vol. 8693, pp. 740–755. Springer, Cham (2014). https://doi.org/10.1007/978-3-319-10602-1_48
26. Hore, A., Ziou, D.: Image quality metrics: PSNR vs. SSIM. In: 2010 20th International Conference on Pattern Recognition, pp. 2366–2369. IEEE, August 2010
27. Wang, Z., Bovik, A.C., Sheikh, H.R., Simoncelli, E.P.: Image quality assessment: from error visibility to structural similarity. IEEE Trans. Image Process. **13**(4), 600–612 (2004)

Abstractive Text Summarization on Templatized Data

C. Jyothi(✉) and M. Supriya(ID)

Department of Computer Science and Engineering, Amrita School of Engineering, Bengaluru,
Amrita Vishwa Vidyapeetham, Bengaluru, India
m_supriya@blr.amrita.edu

Abstract. Abstractive text summarization generates a brief form of an input text from the original source without the sentences being reused by still preserving the meaning and the important information. This could be modelled as a sequence-to-sequence learning by exploiting Recurrent Neural Networks (RNN). Typical RNN models are tough to train owing to the vanishing and exploding gradient complications. 'Long Short-Term Memory Networks (LSTM)' are an answer for such vanishing gradients problem. LSTM based modelling with an attentional mechanism is vital to improve the text summarization application. The key intuition behind attention mechanism is to decide how much attention one needs to pay to every word in the input sequence for generating a word at a particular step. The objective of this paper is to study various attention models and its applicability to text summarization. The intent is to implement Abstractive text summarization with an appropriate attention model using LSTM on a templatized dataset to avoid noise and ambiguity in generating high quality summary.

Keywords: RNN · LSTM · GRU · Text summarization

1 Introduction

We are residing in an era where modern scientific innovations have resulted in the generation of a large volume of data. Every click on the computer, phone or social media produces data. Statistics from the international data corporation (IDC) concludes that, "The total amount of data circulating annually would sprout from 4.4 zetta bytes (10^{21} bytes) in 2013 to 180 zeta bytes in 2025" [1]. With such immense amounts of data available, we need algorithms that will automatically abbreviate long texts to generate precise summaries which can pass the envisioned message. This need brings in the relevance of automatic text summarization. It is the process of shortening a set of data computationally, to create a subset (a summary) that represents the most important or relevant information within the original content [2]. Text Summarization deals with distilling the most important information from a source (or sources) to produce an abridged version for a particular user (or users) and task (or tasks) [3]. If the distillation process is performed with the help of a computer program, then it is called an automatic text summarizing system. This process helps us in faster discovery of relevant information

N. Kumar et al. (Eds.): UBICNET 2021, LNICST 383, pp. 225–239, 2021.
https://doi.org/10.1007/978-3-030-79276-3_17

as well as to consume the information appropriately from the enormous amount of data available online. There are innumerable applications for automatic text summarization [4]. In Telemedicine, summaries of medical records of patients are beneficial to health care providers to analyze and route the case to appropriate health care professionals. Text summaries of recent research papers are generated to know the current trends and innovation in each sector in Science and R&D. Individuals with hearing impairments could take assistance from voice to text summarization to remain updated with the current content. Financial documents such as earning reports and financial news can be summarized with the help of tailor-made summarization systems which can assist analysts to quickly check market indicators.

To address this ever-growing demand of summarizers, various types of summarization systems have been designed. Based on input, text summarization is classified into multi-document and single document systems for summarization. In single document summarization, summary of a single document is created by extracting the important sentences. In multi-document summaries, important sentences are extracted from multiple documents, then they are ranked among each other and important sentences are extracted for which summaries are generated [5]. Based on output, text summarization systems can be categorized as Extractive and abstractive text summarization systems. Extractive summarization is analogous to a highlighter where one reads the document, highlights the important sentences, and writes down those sentences to generate a summary. Abstractive text summarization is analogous to a pen where the document is read, comprehended, and written down as a summary in one's own words. Based on purpose, text summarization is divided in to Generic, Query based and domain specific text summarization systems. Sentences are extracted from multiple documents which are matched with a query and the important sentences are extracted and ranked to generate summaries in Query based text summarization systems. In domain specific text summarization, multiple documents from the domain are pre-processed and saved in a database. When a new document is to be summarized, the important features of the domain are extracted from the database and based on that, the sentences in the document are scored and are ranked and summarized.

In this work, an Abstractive text summarizer is modelled using Sequence to Sequence models with additive attention model on templatized text. This can aid in faster training still giving good accuracy. Seq2Seq model turns one sequence into another sequence with the use of a RNN or more often LSTM or Gated Recurrent Unit (GRU) to avoid the problem of vanishing gradient. The primary components in this model are an Encoder and a Decoder. The input sequence/text is converted into a context vector by an encoder which is passed on as decoder input to another sequence/text based on the context [6]. The encoder-decoder architectural model is generally preferred to convert a given sequence of definite length into a sequence of different length. The encoder converts the input sequence to an internal state which is then passed as input to the decoder to generate another sequence as output. This architecture has various applications such as image captioning, text summarization, machine translation etc. Generally, special categories of recurrent neural networks like LSTM or GRU's are stacked together to be used as encoder and decoder. A recurrent neural network has feedback connections that can be considered as many copies of the same network to produce an output which in turn goes in as input to the next network. This makes RNN the appropriate choice for data involving sequence such as speech and text [7]. But normal RNN's are inefficient to capture long term dependencies in the sequence. The output of the current network will mostly depend only on the immediately preceding information. Hence LSTMs were introduced to overcome this downside of normal RNN's.Text data is messy and deep learning neural networks expects the input to be a fixed length vector. Hence to convert the text into a set of numbers, a tokenizer should be used [8]. Later an appropriate word embedding technique need to be applied to convert the tokenized text into a dense vector. This dense vector will be the input sequence to the encoder. There are some issues like lack of enough vocabulary in training data or sometimes too many trainable parameters due to many words in training data. This is countered using transfer learning approaches where popular models like glove, wordtovec etc. are trained and saved as pre-trained model. These models are used while training other models to save training time and to achieve a rich vocabulary [9]. As the size of the sequence increases such as in text summarization, encoder-decoder architecture which has multiple layers of LSTM stacks become inefficient as the whole context of the entire text need to be consolidated into a single context vector. Hence text summarization problems performed using Seq2Seq models using LSTMs need an additional layer known as attention to capture the context well and generate more accurate summaries. This will enable the decoder to give more importance to some part of the network as compared to others.

Methods for automating text summarization are vital in today's circumstances where there is copiousness of information such as e-mail, news, websites etc. and deficiency of human expertise and time to infer the information. The inspiration to generate summaries are [10] they lessen the interpretation time, makes exploring and selection of paper easier, improve the usefulness of indexing, are less prejudiced compared to human summarizers, can be tailor made and are beneficial in question-answering systems to provide person-alized information, enable abstract services by increasing the number of text documents they can process, do not require expertise in the field as in manual summarization.

This paper is organized in to different sections. Section 2 is separated into concept review and research review where the underlying concepts of attention based on LSTM is discussed in detail along with the mathematical aspects and also the past works in the domain. The technical aspects of the project in terms of various modules and execution details are discussed in Sect. 3. Section 4 presents the outcomes of the project and the paper is concluded in Sect. 5.

2 Literature Review

2.1 Concept Review

LSTM are special types of RNN. The repeating unit in a LSTM network has four layers of neural network unlike one in a normal RNN. These are referred to as gates and are shown in Fig. 1.

The first step in a LSTM network is a forget gate layer whose output is a sigmoid function of the previous hidden state as well as the current input. The weighted sum of previous hidden state and input added with a bias is passed through the sigmoid function,

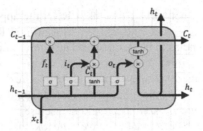

Fig. 1. LSTM network [11]

which outputs a zero or a one as can be seen in Eq. (1). An output of zero means that state needs to be forgotten and one suggests that it needs to be remembered.

$$f(t) = \sigma(W_f \cdot [h_{t-1}, x_t] + b_f) \tag{1}$$

The second step consists of an input gate layer whose output comprise of a sigmoid function of the current input, and the previous hidden state added to a bias like the output of forget layer as can be seen in Eq. (2). It also has a tanh function which generates a list of possible values to be updated in the cell state as shown in Eq. (3). Pointwise multiplication is performed on these outputs in the next stage.

$$i_t = \sigma(W_i \cdot [h_{t-1}, x_t] + b_i) \tag{2}$$

$$\tilde{C}_t = \tanh(W_C \cdot [h_{t-1}, x_t] + b_C) \tag{3}$$

The third step consists of generating the new cell state by point wise addition. The components added are the result of the pointwise multiplication of (i) the previous cell state and forget gate output and (ii) the input gate output and the candidate cell states output as shown in Eq. (4).

$$c_t = f_t * C_{t-1} + i_t * \tilde{C}_t \tag{4}$$

The fourth step consists of generating the output by applying a sigmoid function on the weighted sum of the preceding hidden state and the current input which is then added to a bias as shown in Eq. (5). A pointwise multiplication of the output and the result obtained by applying tanh function on the current cell state is performed to generate the new hidden state as can be seen in Eq. (6).

$$o_t = \sigma(W_o[h_{t-1}, x_t] + b_0) \tag{5}$$

$$h_t = o_t * \tanh(C_t) \tag{6}$$

Using these gates, LSTMs can remember only what is required as it passes the hidden state and cell state from one network to the other. Hence it can avoid vanishing gradient problem to an extent when the sequences are not extremely long and thereby capture their long-term dependencies.

There are two phases in the set-up of Encoder-Decoder architecture in Seq2Seq models namely Training phase and Inference phase. In the training phase of the encoder, initial hidden and cell state are initialized to zero or random numbers and at each step, an input is given to the network which along with the output hidden state of the previous step will try to capture the context. In the last step, the final hidden state is generated to carry the context of the whole input sequence [12]. In the training phase of the decoder, the final state of the encoder will be the initial state for the decoder. The start and end tokens are also appended to the output text before starting to train the model which will enable the decoder to learn where to start and stop. The decoder will be trained using the given training data to generate the output sequence. In the inference phase, the decoder will be given the output of the encoder and the start token which outputs the maximum probability word as the output which along with the current generated hidden state will predict the next word. This process will continue till we reach the end of the sequence.

In the attention model, a dot product operation will be performed on the output vector of each decoder state with the encoder hidden state at each step to generate an attention score for each encoder state. The attention score is then passed through a SoftMax function and an attention distribution will be obtained. This distribution is multiplied with the input sequence and a summation is done to generate the output of the decoder. And this process is repeated with all the decoder inputs to generate a more accurate summary [13].

Additive attention model was first proposed by Bahdanau et. al. [13]. Instead of converting the whole input sequence into a single context vector, this model tries to give importance to specific words in the input sequence with the aid of attention weights. All the hidden states of the encoder including the forward and backward state in a bidirectional LSTM are used to generate a context vector. There after it performs a linear combination of the states of both encoder and decoder, hence the name additive attention.

The attention layer consists of three steps for the generation of alignment score, attention weights and context vector generation. The alignment score tries to evaluate the matching between the input around position "j" and the output position "i". This is calculated using the previous decoder's hidden state, $s_{(i-1)}$ preceding the target word position and the hidden state, h_j of the input sentence. The alignment score is calculated as shown in Eq. (7).

$$e_{ij} = a(s_{i-1}, h_j) \tag{7}$$

The alignment score is passed through a SoftMax function to obtain the attention weight of the corresponding input position 'j' for the output position 'i' as shown in Eq. (8).

$$\alpha_{ij} = \exp(e_{ij}) / \sum_{k=1}^{T_x} \exp(e_{ik}) \tag{8}$$

The corresponding context vector is calculated as per the Eq. (9)

$$C_i = \sum_{j=1}^{T_x} \alpha_{ij} h_j \tag{9}$$

The current hidden state of the decoder is a function of the above context vector, previous hidden state and previous output as can be seen in Eq. (10)

$$s_i = f(s_{i-1}, c_i, y_{i-1}) \tag{10}$$

The current output generated will be a function of the current context vector, current hidden state as well as the previous output as can be seen in Eq. (11)

$$g(y_{i-1}, s_i, c_i) \tag{11}$$

Another type of attention is multiplicative attention. In this method, the attention scores are calculated by matrix multiplication of encoder and decoder states. This improves space efficiency as well as speed. This method was suggested by Luong and hence the name Luong attention [14]. He suggested two types under this category which are global and local attention. Global Attention considers all the words in the input sequence to predict a word in the output sequence. Since all words being considered becomes computationally expensive as the length of the input sequence increases. In Local Attention method, attention is placed only on a subset of the input which can be selected either monotonically or predictively. Since the number of source positions being considered decreases, it is computationally more economical.

The Additive attention method makes use of bidirectional LSTM based encoder's both backward and forward states and combines it with the output of the previous decoding step. This will have a non-stacking uni-directional decoder. The multiplicative attention model considers the hidden states of the topmost LSTM layer in both encoder as well as decoder [14].

2.2 Review of Previous Research Findings

A study conducted by Juan Cao et al. suggests a new methodology for converting table data into textual format using NLDT architecture and then generating summaries on them using encoder-decoder architecture with beam search algorithm in the inference stage [15]. A survey conducted by Wubben et al. compares the different approaches by which we can create a templatized input text to a summarization system to produce more accurate and meaningful summaries [16]. It uses the weatherGov dataset proposed by Lian et al. and modified it to reduce noise and added tags to produce high quality summaries. A study conducted by Lian et al. tries to understand the problem of learning correspondences between the text and summary. In order to handle the high degree of uncertainty, it suggests a model which is generative i.e., it slices the text and plots it to summary. This model has been applied to WeatherGov dataset [17]. Bahdanau et al. tried to improve the quality of output by giving attention weights to each source position in the encoder-decoder architecture rather than encoding the whole input sequence in to a single context vector. This paper revolutionized the way encoder decoder architecture is used and has been widely cited [18]. It also resulted in a separate area of study named attention models and many other attention models are studied thereafter. A model proposed by Nallapati et al. suggests a Hierarchical encoder with hierarchical attention. It also introduces many other novel models to address the challenges in abstractive text summarization and presents a new data set for text summarization [19]. A survey conducted by Shi Tian et al. compares the different approaches to perform Abstractive text summarization. This work proposes an open-source tool kit for generating abstractive summaries and named it as 'Neural Abstractive Text Summarizer' (NATS) [20].

A two-phase model has been proposed by Madhuri et al. [21]. In phase one, it uses a hierarchical fusion of various similarity measures to retrieve the relevant sentences. The hybrid similarity measure incorporates cosine, semantic and word order similarity modules. Phase two consists of a clustering module to remove redundant sentences. This module was realized using DBSCAN agglomerative clustering technique. This model has generated summaries which have shown 86% accuracy. Another model proposed by Ahuja et al. offers means to generate summaries of multiple documents by generation of summaries of single documents and merging them using cosine similarity. The summaries of single documents are generated by extracting features such as Document, Numbers, Noun (proper), Sentence position, and Sentence length (normalized). Total sentence weight is calculated for each sentence and they are extracted to generate multi-document text summary [22]. The heading wise summarizer proposed by Krishnaveni et al. takes from each heading a third of sentences to improve coherence in the final summary generated. The performance improves with the length of the summary and with the number of headings in the document [23]. A model proposed by Siji et al. recommends three level of scoring based on synonym, similarity, and context. But the drawback of the model is that it can't reformat text [24]. A non-dominated genetic algorithm II-based summarizer proposed by Manju Priya et al. uses four different scores namely coherence, thematic, relevance and sentence length to evaluate the significance of sentences [25]. However, this summarizer lacks contextual and semantic information required for better results. A hybrid approach is being proposed by Veena et al. using singular value decomposition and named entity recognition to create a probabilistic model called a Belief Network. This model generates new sentences using subject, verb, and Object. But this model does not include co-reference resolution which can improve the quality of generated sentences [26]. A news summarization system proposed by Karumudi et al. suggests crawling of multiple websites and storing it in a database. It looks for details like person, organization, and location to perform information retrieval of the related news article. This system has the disadvantage of storage access time as well as text processing time delays [27]. A trigraph based extractive text summarization system has been proposed by Devika et al. to extract relevant trigrams in the document. However, in this system, centrality of nodes is only a local measure using indegree and outdegree of each node [28]. A survey conducted by Varalakshmi et al. compares the different approaches to perform extractive text summarization [29]. It recommends swarm intelligence or genetic algorithm technique to extract the best summaries. It also proposes coherence graph to improve coherency between the sentences as well as removing redundant sentences.

Based on the study of these models we can conclude that coherency and redundancy in sentences need to be improved in summaries generated using extractive methods. It is important to know the context of the words in a document as well as in all other documents being considered for summary. Also, many of the previous work in summarization has been extractive. This technique identifies significant sentences, passages in the original document, and replicates them as summary. People on the contrary, tend to reword the original story in their own words. Per se, human summaries are abstractive in nature and hardly ever consist of replication of existing sentences from the document. The existing LSTM based work has used DUC and CNN data set and has the challenge of giving out of vocabulary words (UNK). The weatherGov dataset is templatized to reduce noise and has a smaller vocabulary and can hence obtain better results.

This paper discusses an abstractive text summarization system based on additive attention model on a templatized dataset and the details are discussed in Sect. 3.

3   System Design and Implementation

The proposed problem can be modeled using Deep learning and Natural Language Processing techniques. This can be implemented in python programming language with a suitable integrated development environment (IDE) such as Jupyter Notebook with necessary libraries including TensorFlow, NLTK, Keras, pandas, matplotlib etc.

To implement the problem defined above we need a high computational power runtime environment such as graphic processing unit (GPU). Google colab provides this runtime which can be used up to 12 h continuously in a session. The various components of the system and the implementation details are shown in Fig. 2.

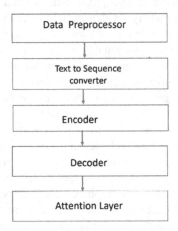

Fig. 2. High Level Design

The data pre-processing module.

a. Removes unwanted tags
b. Removes special characters
c. Removes stop words from the text
d. Drops rows without any entries

The Text to sequence converter

- Converts input text to a sequence of numbers
- Removes those sequence in the decoder output training and validation data which have only <sos> and <eos> tokens

The Encoder is designed with five layers of bidirectional LSTMs stacked on top of each other each with a recurrent drop out and drop out of 0.4. The pretrained Glove embedding was used to convert the tokenized input to the encoder into vectors of 100 dimensions [30]. The decoder is made up of only one unidirectional LSTM layer which has been fed with the tokenized and embedded summaries during training each of which are 100 dimensional vectors. This makes use of pre-trained glove 100 model to reduce training time.

The attention layer used in the model is a custom off the shelf additive attention based on Bahdanau's model. This layer is fed by the encoder and decoder outputs which generates the output of the attention layer. This output is concatenated with the decoder output and then fed to a dense time distributed layer with SoftMax activation function to generate the final outputs of the decoder layer.

The above-mentioned model with encoder decoder architecture based on attention is compiled with optimizer as 'rmsprop' to enable faster convergence and 'sparse categorical cross entropy' as loss function since the output is a vector of numbers. This has been fit in to the training data obtained by splitting the dataset in the ratio 9:1 for training and validation using scikit-learn.

The model was trained on the WeatherGov dataset on a GPU machine. The encoder inputs, the output of the last stage of the encoder and the corresponding hidden state and cell state has been used to create a feature vector. Similarly, a decoder model is set up with corresponding decoder values as in the encoder and an attention model. The given input sequence is then fed into the encoder model to generate the corresponding hidden states. These hidden states along with an <sos> token that acts as the initial decoder input is fed into the corresponding decoder model to generate the next word in the sequence. This sequence is again fed into the decoder model along with hidden state of encoder and this process continues till we reach the <eos> token or maximum summary length. Here the next word is selected based on max probability.

4 Results

WeatherGov dataset consists of 29528 records containing 36 records of weather data paired with natural language forecast with average summary length 28.7 for 3753 cities in the USA over three days. The total vocabulary of this dataset is less than 400. The plot of the length of summary versus number of summaries and length of text versus number of texts is depicted in Fig. 3. It can be inferred that around 96.5% of the summaries are below a length of 60 and 94.2% of the texts are below a length of 120. Hence the max lengths for the training data were set accordingly.

The proposed model was first trained on a GPU with 25000 texts from the WeatherGov dataset. The keras model as explained in Sect. 3 with two stacked bi directional LSTMs in encoder and single unidirectional LSTM in the decoder was set up. Each of these LSTM layer has its embedding layer to create a vector. An off the shelf attention

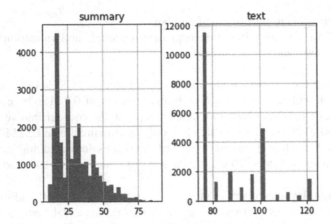

Fig. 3. Length distribution of text and summaries in WeatherGov

model was also included as one layer. In the output side it is converted to a time distributed layer to get the summaries. It was noticed that the number of trainable parameters were around 875,545 which is comparatively less owing to the use of templatized data. The above discussed information has been captured from the model set up using Keras in Fig. 4.

```
Model: "model"
_____
Layer (type)                   Output Shape           Param #    Connected to
=================================================================================
input_1 (InputLayer)           [(None, 120)]          0
_____
embedding (Embedding)          (None, 120, 100)       14000      input_1[0][0]
_____
input_2 (InputLayer)           [(None, None)]         0
_____
bidirectional (Bidirectional)  [(None, 120, 256), (   234496     embedding[0][0]
_____
embedding_1 (Embedding)        (None, None, 100)      30900      input_2[0][0]
_____
concatenate (Concatenate)      (None, 256)            0          bidirectional[0][1]
                                                                 bidirectional[0][3]
_____
concatenate_1 (Concatenate)    (None, 256)            0          bidirectional[0][2]
                                                                 bidirectional[0][4]
_____
lstm_1 (LSTM)                  [(None, None, 256),    365568     embedding_1[0][0]
                                                                 concatenate[0][0]
                                                                 concatenate_1[0][0]
_____
attention_layer (AttentionLayer ((None, None, 256),   131328     bidirectional[0][0]
                                                                 lstm_1[0][0]
_____
concat_layer (Concatenate)     (None, None, 512)      0          lstm_1[0][0]
                                                                 attention_layer[0][0]
_____
time_distributed (TimeDistribut (None, None, 281)     144153     concat_layer[0][0]
=================================================================================
Total params: 920,445
Trainable params: 875,545
Non-trainable params: 44,900
```

Fig. 4. The Keras model summary of proposed system

The model was trained for 10 epochs and as the number of the epoch increases the loss is decreasing and the accuracy is increasing. Between the ninth and tenth epochs there is no significant improvement in the loss and accuracy which shows that the model has converged and training can be stopped. After training for ten epochs, a validation accuracy of 94.53% and a validation loss of 0.16 was obtained. The results of training during various epochs were captured in Fig. 5.

```
Epoch 1/10
322/322 [==============================] - 1555s 5s/step - loss: 0.8702 - accuracy: 0.8044 - val_loss: 0.3950 - val_accuracy: 0.8824
Epoch 2/10
322/322 [==============================] - 1594s 5s/step - loss: 0.3586 - accuracy: 0.8920 - val_loss: 0.3100 - val_accuracy: 0.9048
Epoch 3/10
322/322 [==============================] - 1576s 5s/step - loss: 0.2998 - accuracy: 0.9073 - val_loss: 0.2646 - val_accuracy: 0.9166
Epoch 4/10
322/322 [==============================] - 1573s 5s/step - loss: 0.2619 - accuracy: 0.9172 - val_loss: 0.2387 - val_accuracy: 0.9232
Epoch 5/10
322/322 [==============================] - 1571s 5s/step - loss: 0.2362 - accuracy: 0.9242 - val_loss: 0.2150 - val_accuracy: 0.9298
Epoch 6/10
322/322 [==============================] - 1563s 5s/step - loss: 0.2168 - accuracy: 0.9295 - val_loss: 0.1978 - val_accuracy: 0.9353
Epoch 7/10
322/322 [==============================] - 1575s 5s/step - loss: 0.2002 - accuracy: 0.9343 - val_loss: 0.1829 - val_accuracy: 0.9393
Epoch 8/10
322/322 [==============================] - 1543s 5s/step - loss: 0.1879 - accuracy: 0.9377 - val_loss: 0.1749 - val_accuracy: 0.9407
Epoch 9/10
322/322 [==============================] - 1545s 5s/step - loss: 0.1780 - accuracy: 0.9402 - val_loss: 0.1652 - val_accuracy: 0.9445
Epoch 10/10
322/322 [==============================] - 1542s 5s/step - loss: 0.1701 - accuracy: 0.9426 - val_loss: 0.1600 - val_accuracy: 0.9453
```

Fig. 5. Training results after various Epochs

The plot of train and test validation loss versus the number of epochs is as shown in Fig. 6. As can be noticed from the figure the training loss and testing loss are very close at around 10 epochs. This means that model has been trained well and hence training can be stopped at 10 epochs.

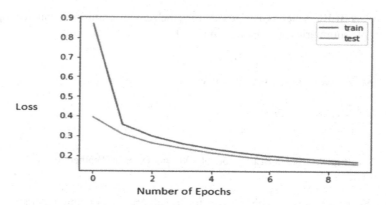

Fig. 6. Train/Test Loss Versus Epochs of WeatherGov

The above model can now be used to predict summaries. A few sample summaries predicted by the model along with the text and the human written summaries used as reference to train the model is shown in Fig. 7.

Metrics for evaluation of a text summarization system include Rouge score and BLEU score. BLEU score which can be expanded as The Bilingual Evaluation understudy score evaluates a machine generated summary against a human summary. If there

```
Text: windchill time 17 30 min 0 mean 34 max 43 windspeed time 17 30 min 10 mean 14 max 18 mode bucket 0 20 2 10 20 win
ddir time 17 30 mode w gust time 17 30 min 0 mean 0 max 0 skycover time 17 30 mode bucket 0 100 4 50 75 skycover time 1
7 21 mode bucket 0 100 4 50 75 skycover time 17 26 mode bucket 0 100 4 75 100 skycover time 21 30 mode bucket 0 100 4 7
5 100 skycover time 26 30 mode bucket 0 100 4 50 75 precippotential time 17 30 min 13 mean 18 max 25 rainchance time 17
21 mode chc rainchance time 17 26 mode chc
Original summary: a 30 percent chance of rain or drizzle before 1am areas of fog before 1am otherwise mostly cloudy wit
h a low around 34 west wind between 10 and 15 mph
Predicted summary:  a chance of showers before midnight mostly cloudy with a low around 49 breezy with a south wind bet
ween 15 and 20 mph chance of precipitation is 30 new rainfall amounts of less than a tenth of an inch possible


Text: temperature time 6 21 min 55 mean 71 max 82 windchill time 6 21 min 0 mean 0 max 0 windspeed time 6 21 min 3 mean
8 max 14 mode bucket 0 20 2 0 10 winddir time 6 21 mode sse gust time 6 21 min 0 mean 7 max 20 skycover time 6 21 mode
bucket 0 100 4 50 75 skycover time 6 9 mode bucket 0 100 4 50 75 skycover time 6 13 mode bucket 0 100 4 50 75 skycover
time 9 21 mode bucket 0 100 4 50 75 skycover time 13 21 mode bucket 0 100 4 50 75 precippotential time 6 21 min 10 mean
10 max 10
Original summary: partly sunny with a high near 82 light wind becoming south between 11 and 14 mph winds could gust as
high as 20 mph
Predicted summary:  partly sunny with a high near 81 south wind between 5 and 15 mph with gusts as high as 20 mph
```

Fig. 7. Sample Summaries of WeatherGov

is a high degree of match, then it assigns a high score. It is a metric which is popular, inexpensive, and automated. This will not check for grammar mistakes or intelligibility. The sentences are compared with the reference sentences and scores are assigned to individual sentences. This is then normalized with the corpus length [31]. The BLEU score obtained for 200 summaries are captured in Fig. 8.

```
The bleu score is: 0.5760084165386521
```

Fig. 8. BLEU score of 200 summaries in WeatherGov

The corpus BLEU which is used to calculate the score for multiple sentences applied on 100 sample summaries are captured in Fig. 9.

```
nltk.translate.bleu_score.corpus_bleu(GOLD, SUMM)

9.372407013347064e-232
```

Fig. 9. Corpus BLEU score of 100 summaries in WeatherGov

Rouge Score which is the abbreviation for Recall oriented gisting under study score calculates the similarity between a generated summary to a reference summary [32]. The rouge score for 100 samples is captured in Fig. 10. Each summary has a rouge1, rouge2 and rougel score associated with it. Each of these has a separate value for precision, recall and f-score linked with it.These information captured for a few sample summaries have been shown in Fig. 10.

```
Predicted summary:  a chance of showers mainly before noon mostly cloudy with a high near 62 southwest wind between 5 an
d 8 mph chance of precipitation is 30 new rainfall amounts of less than a tenth of an inch possible
[{'rouge-1': {'f': 0.5555555505709877, 'p': 0.5882352941176471, 'r': 0.5263157894736842}, 'rouge-2': {'f': 0.34285713787
346944, 'p': 0.36363636363636365, 'r': 0.32432432432432434}, 'rouge-l': {'f': 0.593749995, 'p': 0.59375, 'r': 0.59375}}]
\n
Original summary: occasional rain mainly after 3am low around 42 south wind between 11 and 16 mph chance of precipitatio
n is 80 new rainfall amounts between a quarter and half of an inch possible
Predicted summary: occasional rain mainly after 3am low around 43 south wind between 11 and 14 mph chance of precipitat
ion is 80 new rainfall amounts between a quarter and half of an inch possible
[{'rouge-1': {'f': 0.937499995, 'p': 0.9375, 'r': 0.9375}, 'rouge-2': {'f': 0.8709677369354839, 'p': 0.8709677419354839,
'r': 0.8709677419354839}, 'rouge-l': {'f': 0.9310344777586208, 'p': 0.9310344827586207, 'r': 0.9310344827586207}}]
\n
Original summary: occasional rain after 3am low around 43 south wind between 10 and 13 mph chance of precipitation is 80
new rainfall amounts between a quarter and half of an inch possible
Predicted summary: occasional rain after 3am low around 43 south wind between 10 and 13 mph chance of precipitation is
80 new rainfall amounts between a quarter and half of an inch possible
[{'rouge-1': {'f': 0.999999995, 'p': 1.0, 'r': 1.0}, 'rouge-2': {'f': 0.999999995, 'p': 1.0, 'r': 1.0}, 'rouge-l': {'f':
0.999999995, 'p': 1.0, 'r': 1.0}}]
\n
Original summary: scattered showers mainly before 10am cloudy then gradually becoming mostly sunny with a high near 52 n
orth wind around 7 mph chance of precipitation is 50 new rainfall amounts of less than a tenth of an inch possible
Predicted summary:  a 30 percent chance of showers before noon mostly cloudy with a high near 59 north wind between 3 an
d 7 mph
[{'rouge-1': {'f': 0.4999999953555556, 'p': 0.39473684210526316, 'r': 0.6818181818181818}, 'rouge-2': {'f': 0.2068965471
0463742, 'p': 0.16216216216216217, 'r': 0.2857142857142857}, 'rouge-l': {'f': 0.39285713816964285, 'p': 0.31428571428571
43, 'r': 0.5238095238095238}}]
\n
Original summary: sunny with a high near 33 northwest wind between 8 and 11 mph
Predicted summary:  sunny with a high near 34 northwest wind between 8 and 11 mph
[{'rouge-1': {'f': 0.9230769180769233, 'p': 0.9230769230769231, 'r': 0.9230769230769231}, 'rouge-2': {'f': 0.83333332833
33335, 'p': 0.8333333333333334, 'r': 0.8333333333333334}, 'rouge-l': {'f': 0.9230769180769233, 'p': 0.9230769230769231,
'r': 0.9230769230769231}}]
```

Fig. 10. ROUGE score of 100 summaries in WeatherGov

5 Conclusion

The proposed Seq2seq model-based text summarization system for generating abstractive text summaries was implemented. The encoder and decoder were based on LSTM networks and additive attention was included to improve the quality of summaries and the results were verified using BLEU score and ROUGE metrics. As the proposed system was trained on a templatized data set, the vocabulary of the training data reduced drastically and resulted in faster training and better accuracy. Further studies can be conducted in templatizing the text before training to achieve better scores in abstractive text summarization, especially in domain specific text summarization systems such as the one used in the proposed model. In future, beam search algorithm can be used to select the output sequences to improve the quality of summaries. The repetition of words can be avoided using coverage mechanism. The pointer generator networks can be used to improve the quality which make use of a combination of extractive and abstractive summarization techniques to reduce the training time as well. The trained model can be saved as a hierarchical data format-5(h5) file which is a multidimensional array of scientific data file. This can be deployed using python flask framework in cloud to predict summaries of unseen text.

References

1. https://www.idc.com/
2. https://en.wikipedia.org/wiki/Automatic_summarization
3. Advances in Automatic Text Summarization (The MIT Press) Abridged Edition, Inderjeet Mani

4. https://www.quora.com/What-are-the-real-world-applications-of-automatic-text-summarization
5. https://www.kdnuggets.com/2019/01/approaches-text-summarization-overview.html
6. https://www.analyticsvidhya.com/blog/2020/08/a-simple-introduction-to-sequence-to-sequence-models/
7. https://stanford.edu/~shervine/teaching/cs-230/cheatsheet-recurrent-neural-networks
8. https://keras.io/examples/nlp/pretrained_word_embeddings/
9. https://machinelearningmastery.com/prepare-text-data-deep-learning-keras/
10. https://machinelearningmastery.com/gentle-introduction-text-summarization/
11. https://colah.github.io/posts/2015-08-Understanding-LSTMs/
12. https://www.analyticsvidhya.com/blog/2019/06/comprehensive-guide-text-summarization-using-deep-learning-python/
13. https://www.analyticsvidhya.com/blog/2019/11/comprehensive-guide-attention-mechanism-deep-learning/
14. https://blog.floydhub.com/attention-mechanism/
15. Cao, J.: Generating natural language descriptions from tables. IEEE Access **8**, 46206–46216 (2020). https://doi.org/10.1109/ACCESS.2020.2979115
16. van der Lee, C., Krahmer, E., Wubben, S.: Automated learning of templates for data-to-text generation: comparing rule-based, statistical and neural methods. In: Proceedings of the 11th International Conference on Natural Language Generation, pp. 35–45. Tilburg University, The Netherlands. Association for Computational Linguistics (2018)
17. Liang, P., Jordan, M., Klein, D.: Learning semantic correspondences with less supervision. In: Association of Computational Linguistics (2009)
18. Bahdanau, D., Cho, K., Bengio, Y.: Neural machine translation by jointly learning to align and translate. CoRR, abs/1409.0473 (2014)
19. Nallapati, R., Zhou, B., Gulcehre, C., Xiang, B.: Abstractive text summarization using sequence-to-sequence RNNs and beyond. CoNLL (2016)
20. Shi, T., Keneshloo, Y., Ramakrishnan, N., Reddy, C.K.: Neural abstractive text summarization with sequence-to-sequence models. arXiv abs/1812.02303 (2018)
21. Madhuri Chandu, G.V., Premkumar, A., Susmitha K, S., Sampath, N.: Extractive approach for query based text summarization. In: 2019 International Conference on Issues and Challenges in Intelligent Computing Techniques (ICICT), Ghaziabad, India, pp. 1–5 (2019). https://doi.org/10.1109/ICICT46931.2019.8977708
22. Ahuja, R., Anand, W.: Multi-document text summarization using sentence extraction. In: Dash, S.S., Vijayakumar, K., Panigrahi, B.K., Das, S. (eds.) Artificial Intelligence and Evolutionary Computations in Engineering Systems. AISC, vol. 517, pp. 235–242. Springer, Singapore (2017). https://doi.org/10.1007/978-981-10-3174-8_21
23. Krishnaveni, P., Balasundaram, S.R.: Automatic text summarization by local scoring and ranking for improving coherence. In: 2017 International Conference on Computing Methodologies and Communication (ICCMC), Erode (2017), pp. 59–64 (2017). https://doi.org/10.1109/ICCMC.2017.8282539
24. Rani, S.S., Sreejith, K., Sanker, A.: A hybrid approach for automatic document summarization. In: 2017 International Conference on Advances in Computing, Communications and Informatics (ICACCI), Udupi, pp. 663–669 (2017). https://doi.org/10.1109/ICACCI.2017.8125917
25. Manju Priya, A.R., Gupta, D.: Extractive single document summarization using NSGA-II. In: Thampi, S.M., et al. (eds.) Intelligent Systems, Technologies and Applications. AISC, vol. 1148, pp. 197–211. Springer, Singapore (2020). https://doi.org/10.1007/978-981-15-3914-5_15

26. Veena, G., Gupta, D., Jaganadh, J., Nithya Sreekumar, S.: A graph based conceptual mining model for abstractive text summarization. Indian J. Sci. Technol. **9**(S1) (2017). https://doi.org/10.17485/ijst/2016/v9is1/99876. ISSN 0974-6846

27. Karumudi, G.V.N.S.K., Sathyajit, R., Harikumar, S.: Information retrieval and processing system for news articles in English. In: 2019 9th International Conference on Advances in Computing and Communication (ICACC), Kochi, India, pp. 79–85 (2019). https://doi.org/10.1109/ICACC48162.2019.8986223

28. Raj, D., Geetha, M.: A trigraph based centrality approach towards text summarization. In: 2018 International Conference on Communication and Signal Processing (ICCSP), Chennai, pp. 0796–0801 (2018). https://doi.org/10.1109/ICCSP.2018.8524528

29. Varalakshmi K, P.N., Kallimani, J.S.: Survey on extractive text summarization methods with multi-document datasets. In: 2018 International Conference, on Advances in Computing, Communications and Informatics (ICACCI), Bangalore, pp. 2113–2119 (2018). https://doi.org/10.1109/ICACCI.2018.8554768

30. Pennington, J., Socher, R., Manning, C.: Glove: global vectors for word representation. In: Proceedings of EMNLP, pp. 1532–1543 (2014)

31. https://en.wikipedia.org/wiki/ROUGE_(metric)

32. https://machinelearningmastery.com/calculate-bleu-score-for-text-python/

Author Index

Achuthan, Krishnashree   131
Alam, Md Raiyan   3
Anand, M.   79, 91
Apoorva, Tirumalasetty Sri Sai   56
Arefin, Md. Taslim   3

Balachander, S.   116
Belwal, Meena   105
Binu, Sumitra   187

Devi, M. Nirmala   116
Diallo, Thierno Ahmadou   26
Ding, Zhiguo   66

Evan, Nawshad Ahmad   3

George, Jossy   187
Gouda, Krushna Chandra   131

Hariharan, B. Varshin   116

Jyothi, C.   225

Kalyur, Sesha   150
Kochuvila, Sreeja   198
Kolekar, Uttam D.   16
Kolusu, Pawan Tej   79, 91
Kumar, Navin   16, 56
Kurup, Dhanesh G.   45

Nagaraja, G. S.   150
Nair, Akhil M.   187

Pala, Sreenivasulu   45
Paneerselvam, Surekha   210

Ramakotti, Raksha   210
Ramesh, T. K.   66
Rathina Raj, Vishnu Kanthan   105
Reddy, Sunku Dharahas   198

Sankaran, Sriram   131
Sapkale, Pallavi   16
Seshadrinath, Mukund   131
Shaji, Anchana   187
Shinu, M. R.   169
Shivappa, Himesh   131
Singh Samant, Yogesh Chandra   116
Srinivas, C. G. Prahalad   116
Sumathi, S.   66
Sunku Mohan, Vamshi   131
Supriya, M.   169, 225

Thanmayi A, V. Lakshmi   198

Uddin, Md. Raihan   3

Printed in the United States
by Baker & Taylor Publisher Services